SCIENTIFIC AMERICAN EXPLORES BIG IDEAS

The Science of Stress

The Editors of *Scientific American*

SCIENTIFIC AMERICAN EDUCATIONAL PUBLISHING

Published in 2024 by Scientific American Educational Publishing
in association with **The Rosen Publishing Group**
2544 Clinton Street, Buffalo NY 14224

Contains material from Scientific American®, a division of Springer Nature America, Inc., reprinted by permission, as well as original material from The Rosen Publishing Group®.

First Edition

Scientific American
Lisa Pallatroni: Project Editor

Rosen Publishing
Erica Grove: Compiling Editor
Michael Moy: Senior Graphic Designer

Cataloging-in-Publication Data
Names: Scientific American, Inc.
Title: The science of stress / edited by the Scientific American Editors.
Description: First Edition. | New York : Scientific American Educational Publishing, 2024. | Series: Scientific American explores big ideas | Includes bibliographical references and index.
Identifiers: ISBN 9781725350243 (pbk.) | ISBN 9781725350250 (library bound) | ISBN 9781725350267 (ebook)
Subjects: LCSH: Stress (Physiology)–Juvenile literature. | Stress (Psychology)- -Juvenile literature. | Stress management–Juvenile literature.
Classification: LCC QP82.2.S8 S395 2024 | DDC 305.2350973–dc23

Manufactured in the United States of America
Websites listed were live at the time of publication.

Cover: Evgeny Atamanenko/Shutterstock.com

CPSIA Compliance Information: Batch # CWSA24.
For Further Information contact Rosen Publishing at 1-800-237-9932.

CONTENTS

INTRODUCTION

To be human is to experience stress from time to time. Certain situations simply trigger that "fight or flight" response that raises your heart rate, makes your muscles tense, and causes you to sweat. It can occur while you're taking a difficult test, getting ready to ask someone on a date, or fighting with your best friend. It can also be caused by more serious issues, such as the death of a loved one, financial struggles, or serious illness. Regardless of the situation, it is clear that stress can have significant impacts on both the mind and body.

While stress is a natural response to some situations and can even be beneficial in helping us face challenges and stay safe from threats, research shows that chronic stress can have serious negative impacts on mental and physical health. It has been found to cause issues like insomnia, difficulty focusing, depression, heart attacks, strokes, digestive issues, and a weakened immune system. Because of the significant health risks involved in experiencing a large amount of stress, it is no surprise that so much attention has been given to finding ways to prevent and reduce stress.

The articles in this volume explore the various dimensions of stress, from its causes to its effects to ways to cut down on stress. Section 1, "Stress and the Body," looks at the physical effects of both acute and chronic stress. Section 2, "Stress and the Brain," shifts the focus to stress's psychological impacts and the ways in which activity in the brain can lead to stress. Section 3, "Causes of Stress," considers different factors that contribute to stress, both on an individual and societal level. Finally, Section 4, "Stress Management and Resilience," examines techniques for combatting stress and the characteristics that make certain people more resistant to stress. These articles will offer insight on both the science behind stress and the ways in which this knowledge can be applied to your own life.

Section 1: Stress and the Body

Female Stress—a Faster, Stronger Response

By Debra A. Bangasser

Think about the last time you felt stressed. Did your heart rate quicken? Did your breathing get shallow and fast? Maybe your muscles tensed, and you became more alert? The brain prompts all these physiological changes to help us survive in the face of a potentially life-threatening situation. But when this response is activated inappropriately or persistently, it can become dangerous. Indeed, research has linked uncontrolled stress to a wide range of health problems, from heart disease and diabetes to depression and post-traumatic stress disorder (PTSD).

Women are roughly twice as likely as men to suffer from stress-related psychiatric disorders, according to epidemiological analyses. The big question has always been: Why? Some experts argue that cultural factors are at least partly responsible. For instance, women may be more willing than men to seek help for mental illness, making their cases more likely to be counted. But evidence from animal research suggests that biology may also play an important role. Scientists are uncovering telling differences in the ways that male and female brains react and adapt to stress.

This insight has been a long time coming. Historically, scientists studied male animals almost exclusively—even when investigating disorders that seem to occur more frequently in women. One reason is that many researchers feared that fluctuating ovarian hormones would complicate their studies, muddy their data and create a need for more subjects or more time at greater expense. Recent investigations have discredited this line of reasoning—the data collected from female animals are no more variable than those from males—but the male bias in animal research has persisted nonetheless.

To address this issue, the National Institutes of Health launched an effort in 2016 in the U.S. requiring that scientists conducting animal studies include sex as a biological variable in their research by studying both male and female animals. As a result, researchers who study chronic stress stand a much better chance of understanding how it impacts the health of both sexes—work that could lead to more effective, sex-specific treatments for psychological disorders. In fact, some of the most promising new therapies under investigation—including oxytocin for anxiety and ketamine for depression—appear to have very different effects in females than in males.

Female Stress

The animal models scientists use to explore the effects of stress take many forms. Some researchers expose rodents to something stressful—perhaps a brief restraint or a sound they have conditioned the animals to associate with a mild shock—for several days in a row. Others alter levels of stress-related chemicals, such as glucocorticoids or corticotropin-releasing factor (CRF), in the animals' brains via genetic engineering and other techniques. Regardless of the method, many of these manipulations seem to produce faster and stronger reactions in females. We are just beginning to understand why.

It turns out that the most basic cellular processes involved in the stress response differ between the sexes. For example, neuroscientist Georgia E. Hodes, then working in Scott Russo's laboratory at the Icahn School of Medicine at Mount Sinai, conducted a study in which she and her colleagues stressed male and female mice over the course of several weeks. They noticed that it took 21 days to increase anxiety and depressionlike behaviors in male mice but only six days to produce the same reaction in female mice. In search of an explanation, they looked to the nucleus accumbens, a brain region involved in seeking out rewarding and pleasurable activities. Disruption of normal brain signaling in this area is thought to contribute to anhedonia, or an inability to experience pleasure, which is a common symptom in depression and various other stress-related disorders.

Within the nucleus accumbens, Hodes identified sex differences in the regulation of a gene referred to as *Dnmt3a* (DNA methyltransferase 3a). Following a six-day stressor period, it was expressed more in female mice than in males. This gene encodes an enzyme that alters a cell's DNA in such a way that it prevents other genes from being read and used to make proteins. To determine *Dnmt3a*'s role in chronic stress, Hodes removed the gene in the nucleus accumbens of female mice. Without it, the females became more resilient and responded more like male mice. These findings suggest that female mice experience an increase in *Dnmt3a* expression after only a short exposure to stress, which then blocks other proteins that promote stress resilience. Interestingly, researchers are developing drugs that inhibit DNMT enzymes to treat certain cancers. Therefore, similar drugs may be helpful in treating stress-related disorders, particularly in women.

Changes in gene expression are not the only sex differences observed in the brain. When I was a postdoctoral fellow, my mentor, neuroscientist Rita J. Valentino of Children's Hospital of Philadelphia, and I discovered sex differences in receptors that respond to CRF, a hormone that helps to kick-start the body's biochemical response to stress. Although there are CRF receptors in many brain regions, we focused on the locus coeruleus, a structure responsible for changing our levels of arousal from sleepy to fully awake. During a stressful event, CRF floods the locus coeruleus, where it binds to CRF receptors to keep an animal on high alert. Typically these receptors sit on the outside surface of brain cells, waiting for a CRF signal. As CRF levels rise, however, the receptors migrate from the cell membrane to its interior, effectively going off-line. This process is thought to protect brain cells from becoming overactivated.

We discovered that in male rodents, CRF receptors withdrew inside neurons after exposure to a standard stressor. They also withdrew in male rodents that were genetically modified to overexpress CRF. In female rodents, though, the receptors lingered on the cell membrane, where they could remain responsive to high CRF levels. The results suggest that CRF may increase arousal and alertness more in females than in males. In some situations, this difference may actually be adaptive: it can be a good thing to stay fully switched on during a stressful occurrence. But overactivation of this system can also lead to hyperarousal, a disruptive state that, in humans, contributes to insomnia, impaired concentration and feeling inappropriately "on edge."

Patients with PTSD and depression can exhibit high levels of CRF and these symptoms of hyperarousal. So if similar sex differences are found with human CRF receptors, it could help explain why women are more likely to suffer from PTSD and depression. This variation may be hard to demonstrate, however. When human brains are being imaged, technical limitations make it difficult to detect molecular changes such as CRF receptor localization. But we have other reasons to believe that women, like female mice, may have a greater sensitivity to CRF: injecting it into the bloodstream causes a greater rise in stress hormones in women than in men.

The Impact of Hormones

Where do these differences come from? Emerging research points to the different complement of genes males and females are born with—as well as hormonal surges in the womb and during puberty, both of which can permanently alter the developing brain. Additionally, fluctuating levels of testosterone, estrogens and progesterone can modulate brain function in adults. In my laboratory at Temple University, we assessed the role that these circulating hormones play in regulating rodents' behavioral responses to high levels of CRF by looking at compulsive grooming.

When a rat compulsively licks its own fur, sometimes to the point of causing baldness, it is thought to reflect a state of high anxiety. The compulsive grooming may be a form of self-soothing. We found that when we injected CRF into rats to induce stress, females engaged in more compulsive grooming than males did. Moreover, the amount of grooming changed over the course of a female's estrous cycle, which is similar to the human menstrual cycle but lasts only four to five days. At the phase of the cycle when ovarian hormones—including estrogens and progesterone—peaked, CRF triggered even more compulsive grooming, suggesting that the hormones somehow amplify CRF's effects.

Behavioral responses such as grooming draw on many brain regions. So in trying to account for the grooming differences between the male and female rats, Kimberly Wiersielis, then a graduate student in my lab, speculated that CRF might activate different circuits in their brains. To test this idea, she examined brain slices for cFos, a protein expressed only when brain cells are switched on. Then we statistically compared cell-activation patterns. Our analysis revealed that CRF activated diverse brain regions in both sexes, but the patterns were different—and especially so between males and females at the point during estrus when females have the highest levels of estrogens and progesterone.

A Sex-Based Rx

These sex differences are not minor. As we search for better remedies for stress-related psychiatric conditions, it is vital that we take them into consideration. To date, potential therapies are most often screened in male rodents. But the same compounds can have very different effects in females. For example, neuroscientists Brian Trainor and Michael Q. Steinman, then both at the University of California, Davis, tested one putative therapy, oxytocin, in both male and female mice. Because this hormone promotes social bonding in mammals, scientists have speculated that administering it to people in the form of a nose spray might reduce social anxiety and avoidance, as well as improve deficits in processing social cues, a problem that is seen in some people who have stress-related psychiatric disorders. Their group found that intranasal oxytocin did indeed reduce anxiety in male mice, but it made female mice even more anxious under certain conditions. We need to make sure that oxytocin sprays will not cause similar adverse side effects in women.

Ketamine provides another example. This medication, normally used as an anesthetic, blocks the *N*-methyl-D-aspartate receptor, a protein that can regulate many processes, including aspects of the stress response. It has generated a lot of excitement as a potential therapy for depression because, unlike traditional antidepressants that can take weeks to provide any benefit, low doses of ketamine reduce symptoms fast, sometimes after a single infusion. Unfortunately, high doses of ketamine can induce delirium, hallucinations and an "out of body" experience (which is why it is also a popular recreational drug).

Researchers have been investigating ketamine in male animals to develop more targeted therapies for depression. In 2013, though, neuroscientists Mohamed Kabbaj of the Florida State University College of Medicine and Nicole Carrier, then a graduate student in his lab, also tested ketamine in female rats. They found that it took less ketamine to relieve depressionlike symptoms in the females and that it was not just a matter of differences in body mass. Instead a different biological mechanism seemed to be at work. If these same differences exist in humans, researchers working on new drugs with ketaminelike properties may need to develop separate therapies for men and women. Researchers continue to evaluate ketamine for its long-term safety and effectiveness, and these studies may well reveal that, like female rats, women should receive a lower dose. (In 2019 the Food and Drug Administration approved a ketaminelike drug, called esketamine, for use against treatment-resistant depression.)

We do not know why males and females would have evolved different biological responses to stress. One possibility is that a female animal protecting vulnerable young might be well served by being able to maintain a heightened state of alertness and awareness of her surroundings—as would her babies. That advantage might outweigh the downstream risks associated with increased stress sensitivity, such as greater vulnerability to depression and anxiety disorders. Studies are needed, however, to investigate this idea.

As we commit to including female animals in research, we stand to learn more not only about disorders that are more common in women but also about conditions that are more often diagnosed in men, such as autism and attention-deficit hyperactivity disorder. In this case, looking at brain differences in females might reveal factors that confer greater resilience. Better-targeted treatments for both sexes could follow.

It is becoming increasingly clear that the traditional approach of studying only male animals is flawed. In fact, our past reliance on male animals in drug-development studies may explain, at least in part, why women report more adverse drug reactions than men do. The practice may also have kept drugs that would be highly effective in females from ever making it to market. The good news is that thanks to new efforts that promote the study of both male and female animals, change is underway.

About the Author

Debra A. Bangasser is an associate professor at Temple University, where she is also principal investigator of the Neuroendocrinology and Behavior Laboratory.

Gray Hair Can Return to Its Original Color—and Stress Is Involved, of Course

By Diana Kwon

Few harbingers of old age are clearer than the sight of gray hair. As we grow older, black, brown, blonde or red strands lose their youthful hue. Although this may seem like a permanent change, new research reveals that the graying process can be undone—at least temporarily.

Hints that gray hairs could spontaneously regain color have existed as isolated case studies within the scientific literature for decades. In one 1972 paper, the late dermatologist Stanley Comaish reported an encounter with a 38-year-old man who had what he described as a "most unusual feature." Although the vast majority of the individual's hairs were either all black or all white, three strands were light near the ends but dark near the roots. This signaled a reversal in the normal graying process, which begins at the root.

In a study published today in *eLife*, a group of researchers provide the most robust evidence of this phenomenon to date in hair from around a dozen people of various ages, ethnicities and sexes. It also aligns patterns of graying and reversal to periods of stress, which implies that this aging-related process is closely associated with our psychological well-being.

These findings suggest "that there is a window of opportunity during which graying is probably much more reversible than had been thought for a long time," says study co-author Ralf Paus, a dermatologist at the University of Miami.

Around four years ago Martin Picard, a mitochondrial psychobiologist at Columbia University, was pondering the way our cells grow old in a multistep manner in which some of them begin to show signs of aging at much earlier time points than others. This patchwork process, he realized, was clearly visible on our head, where our hairs do not all turn gray at the same time. "It seemed like the hair, in a way, recapitulated what we know happens at the cellular level," Picard says. "Maybe there's something to learn there. Maybe the hairs that turn white first are the more vulnerable or least resilient."

While discussing these ideas with his partner, Picard mentioned something in passing: if one could find a hair that was only partially gray—and then calculate how fast that hair was growing—it might be possible to pinpoint the period in which the hair began aging and thus ask the question of what happened in the individual's life to trigger this change. "I was thinking about this almost as a fictive idea," Picard recalls. Unexpectedly, however, his partner turned to him and said she had seen such two-colored hairs on her head. "She went to the bathroom and actually plucked a couple—that's when this project started," he says.

Picard and his team began searching for others with two-colored hairs through local ads, on social media and by word of mouth. Eventually, they were able to find 14 people—men and women ranging from nine to 65 years old with various ethnic backgrounds (although the majority were white). Those individuals provided both single- and two-colored hair strands from different parts of the body, including the scalp, face and pubic area.

The researchers then developed a technique to digitize and quantify the subtle changes in color, which they dubbed hair pigmentation patterns, along each strand. These patterns revealed something surprising: In 10 of these participants, who were between age nine and 39, some graying hairs regained color. The team also found that this occurred not just on the head but in other bodily regions as well. "When we saw this in pubic hair, we thought, 'Okay, this is real,'" Picard says. "This happens not just in one person or on the head but across the whole body." He adds that because the reversibility only appeared in some hair follicles, however, it is likely limited to specific periods when changes are still able to occur.

Most people start noticing their first gray hairs in their 30s—although some may find them in their late 20s.This period, when graying has just begun, is probably when the process is most reversible, according to Paus. In those with a full head of gray hair, most of the strands have presumably reached a "point of no return," but the possibility remains that some hair follicles may still be malleable to change, he says.

"What was most remarkable was the fact that they were able to show convincingly that, at the individual hair level, graying is actually reversible," says Matt Kaeberlein, a biogerontologist at the University of Washington, who was one of the editors of the new paper but was not involved in the work. "What we're learning is that, not just in hair but in a variety of tissues, the biological changes that happen with age are, in many cases, reversible–this is a nice example of that."

The team also investigated the association between hair graying and psychological stress because prior research hinted that such factors may accelerate the hair's aging process. Anecdotes of such a connection are also visible throughout history: according to legend, the hair of Marie Antoinette, the 18th-century queen of France, turned white overnight just before her execution at the guillotine.

In a small subset of participants, the researchers pinpointed segments in single hairs where color changes occurred in the pigmentation patterns. Then they calculated the times when the change happened using the known average growth rate of human hair: approximately one centimeter per month. These participants also provided a history of the most stressful events they had experienced over the course of a year.

This analysis revealed that the times when graying or reversal occurred corresponded to periods of significant stress or relaxation. In one individual, a 35-year-old man with auburn hair, five strands of hair underwent graying reversal during the same time span, which coincided with a two-week vacation. Another subject, a 30-year-old woman with black hair, had one strand that contained a white segment that corresponded to two months during which she underwent marital separation and relocation–her highest-stress period in the year.

Eva Peters, a psychoneuroimmunologist at the University Hospital of Giessen and Marburg in Germany, who was not involved in this work, says that this is a "very creative and well-conceptualized study." But, she adds, because the number of cases the researchers were able to look at was relatively small—particularly in the stress-related portion of the study—further research is needed to confirm these findings.

For now, the next step is to look more carefully at the link between stress and graying. Picard, Paus and their colleagues are currently putting together a grant to conduct another study that would examine changes in hair and stress levels prospectively—which means tracking participants over a specified period of time rather than asking them to recall life events from the past.

Eventually, Picard says, one could envision hair as a powerful tool to assess the effects of earlier life events on aging—because, much like the rings of a tree, hair provides a kind of physical record of elapsed events. "It's pretty clear that the hair encodes part of your biological history in some way," he says. "Hair grows out of the body, and then it crystallizes into this hard, stable [structure] that holds the memory of your past."

About the Author

Diana Kwon is a freelance journalist who covers health and the life sciences. She is based in Berlin.

The Danger of Stress

By Melinda Wenner

You probably think you're doing everything you can to stay healthy: you get lots of sleep, exercise regularly and try to avoid fried foods. But you may be forgetting one important thing. Relax! Stress has a bigger impact on your health than you might realize, according to research presented yesterday at the annual conference of the American Psychological Association in Boston.

Ohio State University psychologist Janice Kiecolt-Glaser and her partner, Ronald Glaser, an OSU virologist and immunologist, have spent 20-odd years researching how stress affects the immune system, and they have made some startling discoveries. An easy example comes from their work with caregivers, people who look after chronically ailing spouses or parents (no one would argue that this role is quite stressful). In one experiment, Kiecolt-Glaser and her colleagues administered flu vaccines to caregivers and control subjects and compared the numbers of antibodies–proteins involved in immune reactions–that the two groups produced in response. Only 38 percent of the caregivers produced what is considered an adequate antibody response compared to 66 percent of their relaxed counterparts, suggesting that the caregivers' immune systems weren't doing their jobs very well–and that the stress of caregiving ultimately put them at an increased risk of infection.

If stress affects immune responses, then it should also affect how well the body heals itself. In one particularly cringe-worthy study, Kiecolt-Glaser and her colleagues afflicted a group of caregivers with small arm wounds using a tool dermatologists use to perform skin biopsies. The caregivers' wounds took 24 percent longer to heal than wounds that they had afflicted to non-caregivers.

Okay, but what if caregiving isn't an accurate proxy for stress? To explore other stressful situations, Kiecolt-Glaser and colleagues performed another experiment in which they produced tiny lesions in the mouths of–quite appropriately–11 dental students at two different points in time. Once was during their summer vacation, when they were relaxed, and once was during the fall, several days before a difficult exam. The lesions done before the exam took from two to eight days longer to heal than the summer wounds. For some subjects, the exam wounds took nearly twice as long to get better. Ouch.

How Stress Hurts

It might seem counterintuitive, but Kiecolt-Glaser believes that stress makes our immune systems less effective because it actually elicits an immune response itself. Stress, she says, causes the body to release pro-inflammatory cytokines, immune factors that initiate responses against infections. When the body produces these cytokines over long periods of time—for instance, as a result of chronic stress—all sorts of bad things can happen. Not only does it hamper our body's ability to fight infection and heal wounds, but chronic inflammation also increases our risk of heart disease, osteoporosis, and autoimmune diseases including type 2 diabetes.

What's more, because regular stress causes a chronic immune response, it can also increase a person's risk for allergies, which occur when the body elicits a chronic immune response against something that's not really dangerous (like pollen). In her most recent study, announced yesterday, Kiecolt-Glaser found that when people are under lots of stress—for instance, when they are forced to deliver a speech or do difficult math problems on the spot—their allergies worsen over the course of the next day.

I admit, this post may not exactly have the intended effect—worrying about the danger of stress is definitely stressful—but take a deep breath. If you ask me, learning a few relaxation techniques sounds like a more appealing illness-prevention strategy than many other alternatives. (Go ahead and throw away that cod liver oil.) Indeed, the next time you feel guilty about enjoying a lazy Sunday, or taking a day off from work for the heck of it, think again. Your body will thank you for it.

About the Author

Melinda Wenner is a journalist based in New York City.

How Short-Term Stress Boosts Immune Systems

By Ferris Jabr

Firdaus Dhabhar likes to film babies crying when they get their shots, but not for any sadistic reason. He believes that the wailing is a good sign. A Stanford University researcher who studies how stress changes the body, Dhabhar, along with his colleagues, has discovered that stressed-out laboratory mice exhibit more robust immune responses to vaccines than control groups of mice left in peace. Something similar happens to people. In a study of knee surgery patients, for example, Dhabhar found that the anxiety of their impending operations boosted the number of immune cells circulating in their blood. Such studies have convinced Dhabhar that stress does not entirely deserve its bad reputation and that in some situations it can actually improve health.

Dhabhar and his collaborators contrast the benefits of short-term stress with the consequences of chronic stress, which has long been known to suppress the immune system. Then again, chronic stress can also exacerbate allergies, asthma and autoimmune disorders in which the immune system is already overactive. So does stress excite or repress the immune system? Here is where things get frustratingly fuzzy, as they so often do in biology. It turns out that the answer hinges on the situation and the individual. A transient burst of stress tends to activate some parts of the immune system but not others; conversely, chronic stress generally stifles the entire immune system and can make it more likely to attack benign tissues.

In the knee surgery study, patients' immune systems did not all respond equally to anticipation of the operation. Some people showed an agile, adaptive response: the number of immune cells in their bloodstream peaked in the days before the operation, then decreased as those cells migrated to other tissues throughout the body. Other patients had a more sluggish maladaptive response:

their levels of immune cells hardly wavered from baseline. As you might expect, those with adaptive immune responses recovered from surgery faster.

Most likely it will take decades of new research to gain a much deeper understanding of the biological mechanisms behind such individual differences. For now, though, we can at least be sure that it is okay to feel stressed when you get a shot—in fact, it's a good thing.

About the Author

Ferris Jabr is a contributing writer for Scientific American. *He has also written for the* New York Times Magazine, *the* New Yorker *and* Outside.

How Stress Messes With Your Gut

By Carin Leong, Tanya Lewis, Josh Fischman

Tanya Lewis: Hi, this is *Your Health, Quickly*, a *Scientific American* podcast series!

Josh Fischman: We highlight the latest vital health news: discoveries that affect your body and your mind.

Lewis: And we break down the medical research to help you stay healthy.

I'm Tanya Lewis.

Fischman: I'm Josh Fischman.

Lewis: We're *Scientific American*'s senior health editors.

On today's show, we're talking about how the stuff that goes on in your brain is connected to symptoms in your gut. Scientists have found molecular pathways that trigger inflammatory bowel disease flare-ups. And those paths start in the brain, triggered by stress. This raises the possibility of a new type of treatment.

[CLIP: Show theme music]

Fischman: Hey, remember the other day when we were comparing belly aches?

Lewis: Yeah, I do remember that. I was feeling a bit off, but drinking ginger ale really helped settle my stomach. Did your antacid tablets end up helping?

Fischman: Yeah they did. But neither of us was really sick. I mean, we both showed up at work, after all.

Lewis: Yeah, I mean, speak for yourself–I wasn't feeling great. But I definitely wasn't extremely ill.

Fischman: There is another kind of gastrointestinal illness, though, that's in a whole different league. It's inflammatory bowel disease, or IBD. That's bouts of disabling pain, bleeding, weight loss, diarrhea and hospitalizations that sometimes require surgery.

Lewis: That really doesn't sound like fun. I actually know some people with IBD. Is it a super common illness?

Fischman: Unfortunately, it is. About 3 million adults in the U.S. have it.

Lewis: I've heard of Crohn's disease, which can be pretty disabling. Isn't that a type of IBD?

Fischman: Yeah, it is. The other type is called ulcerative colitis.

Lewis: Oh, I've heard of that one too. I think it mainly affects the large intestine.

Fischman: And Crohn's often damages the small intestine, although it can hurt other parts of the GI tract as well.

What makes them both IBD is inflammation. IBD is often linked to an overactive immune system, where the body's own immune cells attack the digestive tract. And it may have a genetic component.

Lewis: Yeah, I've heard that Crohn's is more common among people with a family history of it. But are there any effective treatments?

Fischman: That's usually some variety of immune suppressant such as a steroid drug or a medication that soothes inflammation. But even these controlled cases have periodic flare-ups.

The causes of these flares have been a real mystery. But I've been reading that the reason—or at least one of the main reasons—can be psychological stress.

Lewis: That does make sense. But it seems a little bit obvious that stress triggers a lot of diseases. So what's the new thing here?

Fischman: It's the precise connections. Starting in the brain, researchers traced two different pathways, made up of molecules and cells that kind of bang into one another. And the paths led all the way down to the intestines.

Lewis: So is that kind of like a series of falling dominoes, each one knocking the next one over?

Fischman: Yeah that's a good way of thinking of it.

Lewis: But I just want to be clear: you're not saying that IBD is all in your head, right?

Fischman: No, no. Really no. It's a biological disease tied to those hyperactive immune cells. But...

Christoph Thaiss: Psychological stress will greatly exacerbate the severity of the disease. So stress by itself is not causing the disease, but it's greatly increasing the magnitude of the disease.

Fischman: That's Christoph Thaiss.

Thaiss: I'm an assistant professor of microbiology here at the Perelman School of Medicine at the University of Pennsylvania.

Fischman: Thaiss was one of the scientists who traced out the rows of dominoes, as you put it. And his team found that the first domino to fall with IBD patients was some kind of stressful experience.

Thaiss: If they had broken up with their partner or lost their jobs or lost someone in the family or any other major event in the lives of these patients ... if someone goes through a very stressful period in their lives, it feels like their disease flaring up, but the physician doesn't see microscopic signs of inflammation yet. But then very often what will happen is a few weeks down the road, then there will be a flare. So there is definitely a very strong component of how the brain or the psychological state of the patient can control inflammatory diseases.

Lewis: That is really validating, and I can say from personal experience that that intense stress can affect the body. But it sounds kind of anecdotal. I'm guessing Thaiss did some more scientific studies?

Fischman: Hey, I don't blame you for a little skepticism. There's a lot of loose talk about how stress affects the body, and you're right: Thaiss and his team did go after this more scientifically. First they looked at mice with IBD-like conditions in order to identify the rows of molecular dominoes.

Then they verified that these same dominoes fell down in people. The pattern held true in three different groups of IBD patients. They even went so far as to do colonoscopy exams on some of these people, and saw these molecular signs of inflammation after stressful events.

Lewis: So what's the next domino after someone has a fight with their spouse, for example?

Fischman: That's the release of glucocorticoids, those are hormones that the brain triggers when you feel threatened. And these hormones reach two different kinds of cells in the gut, with two different effects.

Lewis: You mean hormones like cortisol? How does that affect cells in the gut?

Fischman: Well, first, did you know the gut had its own nervous system?

Lewis: Yeah, I think I did. Isn't it called the enteric nervous system? I think it has neurons and supporting cells.

Fischman: So those supporting cells are called glial cells. They do a bunch of different things and one is to signal those hyperactive immune cells, the ones I mentioned earlier, when the body is stressed.

Those cells arrive, kind of like an attacking army, and they hit the lining of the intestines. And bingo, you get inflammation and a flare-up of IBD.

Lewis: And what about the second cell type you mentioned?

Fischman: Those are the enteric neurons. They control the muscles of the intestines and therefore how quickly or slowly food moves through them. Long exposure to glucocorticoids blocks these neurons from developing fully. They kind of stay in an immature state.

And immature neurons aren't able to make muscles squeeze very hard. So food moves very slowly. And IBD patients feel badly bloated or constipated or crampy. It just makes everything worse.

Lewis: That sounds really horrible. But if stress leads to IBD flare-ups, could treating the stress help patients?

Fischman: Thaiss thinks so, although he cautions these are early days. Nobody has rigorously tested interventions like psychotherapy on IBD.

But he believes that stress might affect how well other treatments, like drugs, actually work.

Thaiss: The treatment response of a patient might very strongly depend on psychological factors, which would mean that we should assess the psychological factors and tailor the treatment accordingly.

Lewis: So that's not something doctors already do?

Thaiss: No, it's not a routine practice to basically quantify a patient's level of psychological stress when they're first seen by a physician or when the decision about the treatment is being made. So that's something that we're now following up on.

Fischman: Thaiss also thinks that psychotherapy itself, including stress management techniques, might help.

Thaiss: Stress mitigation strategy, which I think would be potentially even more beneficial than the molecular interventions at the downstream end of the pathway ... it may suggest that we can block this inflammatory pathway altogether by improving a patient's mental state or emotional state.

Lewis: I guess I finally need to do that mindfulness meditation that I've been putting off. But it totally makes sense that treating stress itself could help with physical symptoms. And it seems like a promising place to start.

Do you think it'll work?

Fischman: Well, I hope it does. Like Thaiss said, they really need to test it. But I do have a *gut feeling* there may be something to this.

[CLIP: Show theme music]

Fischman: *Your Health, Quickly* is produced by Tulika Bose, Jeff DelViscio, Kelso Harper and Carin Leong. It's edited by Elah Feder and Alexa Lim. Our music is composed by Dominic Smith.

Lewis: Our show is a part of *Scientific American*'s podcast, *Science, Quickly*. You can subscribe wherever you get your podcasts. If you like the show, give us a rating or review!

And if you have ideas for topics we should cover, send us an email at Yourhealthquickly@sciam.com. That's your health quickly at S-C-I-A-M dot com.

Lewis: I'm Tanya Lewis.

Fischman: I'm Josh Fischman.

Lewis: We'll be back in two weeks. Thanks for listening!

About the Authors

Carin Leong is a multimedia intern producing podcasts and videos at Scientific American.

Tanya Lewis is a senior editor covering health and medicine for Scientific American. *Follow her on Twitter @tanyalewis314.*

Josh Fischman is a senior editor at Scientific American *who covers medicine, biology and science policy. He has written and edited about science and health for* Discover, Science, Earth, *and* U.S. News & World Report.

Fact or Fiction?: Stress Causes Gray Hair

By Coco Ballantyne

Legend has it that Marie Antoinette's hair turned white the night before she was guillotined. Presumably the stress of impending decapitation caused her locks to lose color within hours. Extremely unlikely, scientists say, but stress may play a role in a more gradual graying process.

The first silvery strands usually pop up around age 30 for men and age 35 for women, but graying can begin as early as high school for some and as late as the 50s for others.

Graying begins inside the sunken pits in the scalp called follicles. A typical human head has about 100,000 of these teardrop-shaped cavities, each capable of sprouting several hairs in a lifetime. At the bottom of each follicle is a hair-growing factory where cells work together to assemble colored hair. Keratinocytes (epidermal cells) build the hair from the bottom up, stacking atop one another and eventually dying, leaving behind mostly keratin, a colorless protein that gives hair its texture and strength. (Keratin is also a primary component of nails, the outer layer of skin, animal hooves and claws, even rhinoceros horns.)

As keratinocytes construct hairs, neighboring melanocytes manufacture a pigment called melanin, which is delivered to the keratinocytes in little packages called melanosomes.

Hair melanin comes in two shades—eumelanin (dark brown or black) and pheomelanin (yellow or red)—that combine in different proportions to create a vast array of human hair colors. Hair that has lost most of its melanin is gray; hair that has lost all of this pigment is white.

At any given time, around 80 to 90 percent of the hairs on a person's head are in an active growth phase, which may last anywhere from two to seven years. At the end of this stage, the follicle shrivels, the keratinocytes and melanocytes undergo programmed cell death (apoptosis), and the follicle enters a resting phase, during which the hair falls out.

To begin building a new hair, the follicle factory must be rebuilt. Fresh keratinocytes and melanocytes are recruited from progenitor cells, also called "stem cells," residing at the bottom of the follicle. For unknown reasons, keratinocyte stem cells have a much greater longevity than the melanocyte stem cells, says David Fisher, professor of pediatrics at Harvard Medical School. "It's the gradual depletion of [melanocyte] stem cells that leads to the loss of pigment," he says.

Does stress accelerate this demise of the melanocyte population? "It is not so simple," Fisher says, noting that the process of graying is a multivariable equation. Stress hormones may impact the survival and/or activity of melanocytes, but no clear link has been found between stress and gray hair.

Suspicions–and hypotheses–abound, however. "Graying could be a result of chronic free radical damage," says Ralf Paus, professor of dermatology at the University Hospital Schleswig-Holstein in Lübeck, Germany. Stress hormones produced either systemically or locally (by cells in the follicle) could produce inflammation that drives the production of free radicals–unstable molecules that damage cells–and "it is possible that these free radicals could influence melanin production or induce bleaching of melanin," Paus says.

"There is evidence that local expression of stress hormones mediate the signals instructing melanocytes to deliver melanin to keratinocytes," notes Jennifer Lin, a dermatologist who conducts molecular biology research at the Dana-Farber / Harvard Cancer Center in Boston. "Conceivably, if that signal is disrupted, melanin will not deliver pigment to your hair."

And general practice physicians have observed accelerated graying among patients under stress, says Tyler Cymet, head of family medicine at Sinai Hospital in Baltimore, who conducted a small retrospective study on hair graying among patients at Sinai. "We've seen that people who are stressed two to three years report that they turn gray sooner," he says.

Cymet suspects that going gray is "genetically outlined, but stress and lifestyle give you variation of plus or minus five to 10 years." Blonds often appear to go gray later in life because white strands easily hide in a sea of light hair when in fact those who are likely to have the darkest hair (people of African and Asian ancestry) seem to retain their color longer.

In short, scientists are beginning to gather clues that stress can hasten the graying process, but there is no scientific evidence demonstrating a cause-and-effect relationship.

So what happened to Marie Antoinette? There are at least three possible explanations: She may have suffered from sudden onset of the rare autoimmune disease alopecia areata, which attacks pigmented hairs, causing them to fall out, leaving the white (nonpigmented) strands behind. Or the stress of the situation could have generated a swarm of free radicals in her hair follicles, which traveled along the hair shafts, destroying pigment and creating a bleaching effect. Or maybe she just stopped wearing her wigs—revealing she had gray hair all along.

Does Stress Feed Cancer?

By Katie Moisse

A little stress can do us good–it pushes us to compete and innovate. But chronic stress can increase the risk of diseases such as depression, heart disease and even cancer. Studies have shown that stress might promote cancer indirectly by weakening the immune system's anti-tumor defense or by encouraging new tumor-feeding blood vessels to form. But a new study published April 12 in *The Journal of Clinical Investigation* shows that stress hormones, such as adrenaline, can directly support tumor growth and spread.

For normal cells to thrive in the body, "they need to be attached to their neighbors and their surroundings," says the study's lead author Anil Sood from The University of Texas M. D, Anderson Cancer Center in Houston. Cells that detach from their environment undergo a form of programmed cell death called anoikis. "But cancer cells have come up with way to bypass this effect–they avoid anoikis," Sood says. This allows cancer cells to break off from tumors, spread throughout the body (in blood or other fluid) and form new tumors at distant sites–a process called metastasis. So Sood wondered: Could stress affect anoikis? "It surprised us that this biology hadn't been studied before," he notes. "Stress influences so many normal physiological processes. Why wouldn't it be involved in tumor progression?"

Sood and his team first studied the effects of stress hormones on human ovarian cancer cell anoikis in culture. Cells that were exposed to stress hormones were protected from self destruction–meaning they could survive without being anchored to their surroundings. The stress hormone treatment activated a protein called FAK (focal adhesion kinase), which is known to protect cells from anoikis. Inhibiting FAK reversed the effects.

But real tumors behave differently than cancer cells in vitro, so Sood and his team extended their exciting findings into a mouse model of cancer. After receiving a transplant of ovarian cancer cells, mice were restrained to cause stress. As such, their tumors grow more quickly. Isoproterenol (a drug similar to adrenaline) had the same accelerative effect. The tumor-feeding effects of behaviorally and pharmacologically induced stress, both of which were mediated by FAK, were inhibited by the adrenaline-blocking drug propranolol.

But how closely does the stress caused by restraining a lab animal mimic that experienced by human patients? Sood and colleagues looked at samples from 80 cases of human ovarian cancer grouped according to patient stress using the National Institutes of Health's Center for Epidemiological Studies Depression scale as a surrogate marker. Patient stress (according to the scale), along with elevated stress hormone activity were associated with higher levels of activated FAK, which was in turn linked to faster disease progression.

Ovaries contain high levels of stress hormones compared with other organs, but Sood plans to investigate whether the stressors could still be involved in other types of cancer. He hopes to identify ways to interfere with the tumor-feeding effects of stress hormones either behaviorally or pharmacologically. "Reducing the hormone levels may not be so easy," Sood says. "Blocking the receptors using drugs like beta-blockers or antidepressants may be a better strategy." Teaching patients to manage their stress using cognitive behavioral therapy might also be effective, he adds. "We're really trying to understand the biology. We hope it will help us identify better therapeutic strategies."

How Dad's Stresses Get Passed Along to Offspring

By Esther Landhuis

A stressed-out and traumatized father can leave scars in his children. New research suggests this happens because sperm "learn" paternal experiences via a mysterious mode of intercellular communication in which small blebs break off one cell and fuse with another.

Carrying proteins, lipids and nucleic acids, these particles ejected from a cell act like a postal system that extends to all parts of the body, releasing little packages known as extracellular vesicles. Their contents seem carefully chosen. "The cargo inside the vesicle determines not just where it came from but where it's going and what it's doing when it gets there," says Tracy Bale, a neurobiologist at the University of Maryland School of Medicine.

Preliminary research by Bale and others, announced in November at the annual meeting of the Society for Neuroscience in San Diego, shows how extracellular vesicles can regulate brain circuits and help diagnose neurodegenerative diseases–in addition to altering sperm to disrupt the brain health of resulting offspring.

Striking evidence that harsh conditions affect a man's children came from crop failures and war-ravaged Europe more than a century ago. In those unplanned human experiments, prolonged famine appeared to set off a host of health changes in future generations, including higher cholesterol levels and increased rates of obesity and diabetes. To probe the inheritance of such changes at the cellular level, Bale and co-workers performed a series of mouse experiments.

It is pretty easy to stress out a mouse. Stick one into a tube it cannot wriggle out of, soak its bedding or blast white noise–and stress hormone levels shoot up, much as they do in people worrying about finances or facing incessant pressure at work. Remarkably, the way a mouse physiologically responds to stress looks noticeably different if–months before conception–its father endured a period of stress. Somehow "their brain develops differently than if their dad hadn't experienced that stress," says Chris Morgan, a postdoc in Bale's lab who helped create the mouse model.

The big question is how information about the paternal environment reaches the womb in the first place. After all, Morgan says, the "dad is only in there for one night, perhaps just a few hours." Could his sperm carry memories of prior trauma? The idea seemed reasonable yet controversial. Because DNA is packed so tightly in the nucleus of a sperm cell, "the thought that [the cell] would respond to anything in the environment really boggled people's minds," says Jennifer Chan, a former Ph.D. student in Bale's lab who is now a postdoc at Icahn School of Medicine at Mount Sinai in New York City.

Rather, there must be some other kind of cell whose DNA does react to environmental changes–and that cell, she reasoned, could then relay that information to sperm cells to transmit at fertilization. She focused on a population of cells that interact with developing sperm by releasing molecules that help sperm grow and mature. They also secrete extracellular vesicles–and Chan showed it is these vesicles whose contents fuse with sperm cells, instilling memories of dad's prior stress.

In one set of experiments Chan stressed a group of male mice, let them mate and looked at stress responses in the pups. The clincher was a set of in vitro fertilization–like experiments in which she collected sperm from a male mouse that had never experienced induced stress. Half his sperm went into a lab dish with vesicles previously exposed to stress hormones. The other half was cultured with vesicles that had no contact with stress hormones.

Chan injected sperm cells from each batch into eggs from a nonstressed female, then implanted the fertilized eggs–zygotes–into the same foster mom. The pups from nonstressed zygotes developed normally. Pups from stress-exposed zygotes, however, showed the same abnormal stress response as those whose dads had experienced stress before mating. That showed extracellular vesicles act as the conduit for transmitting paternal stress signals to the offspring, Chan says.

The findings are "novel and of very high impact, especially when we consider the impact of military service or other work environments that can confer high stress," says Robert Rissman, a neuroscientist at the University of California, San Diego, who was not involved with the research. "I think it would be important to better understand the specificity of the effect and how different types of stressors or strength of stressors can modulate this system."

As a first step toward translating the findings to people, Morgan is collaborating with University of Pennsylvania psychiatrist Neill Epperson to track protein and RNA changes in human sperm samples. At the neuroscience meeting, Morgan presented data from a six-month study of 20 undergraduate and graduate students. Each month the participants came in and gave a sperm donation. They also completed a same-day survey asking how stressed they were feeling. Preliminary data suggest just several months after a student reports stress, his sperm shows changes in "small noncoding RNAs"–RNA molecules that do not get translated to protein but instead control which genes get turned on or off.

Analyzing sperm from this group of healthy young men, the researchers plan to build a basic understanding of molecular changes linked with mild stresses such as taking final exams. In the future Bale and colleagues hope to compare these baseline fluctuations with changes induced by more prolonged life stressors such as post-traumatic stress disorder or neurological diseases such as autism and schizophrenia.

The molecular signatures in extracellular vesicles may also help researchers discover new ways to noninvasively diagnose or predict adverse health outcomes in offspring, says Gerlinde Metz, who studies transgenerational inheritance of stress responses at the University of Lethbridge in Alberta and was not involved with the research. If so, the vesicles could become the basis for a pioneering type of stress test.

About the Author

Esther Landhuis is a freelance science and health journalist based in the San Francisco Bay Area. Follow her on Twitter @elandhuis.

Section 2: Stress and the Brain

Everyday Stress Can Shut Down the Brain's Chief Command Center

By Amy Arnsten, Carolyn M. Mazure, Rajita Sinha

The entrance exam to medical school consists of a six-hour fusillade of hundreds of questions that, even with the best preparation, often leaves the test taker discombobulated and anxious. For some would-be physicians, the relentless pressure causes their reasoning abilities to slow or even shut down entirely. The experience—known variously as choking, brain freeze, nerves, jitters, folding, blanking out, the yips or a dozen other descriptive terms—is all too familiar to virtually anyone who has flubbed a speech, bumped up against writer's block or struggled through a lengthy exam.

For decades scientists thought they understood what happens in the brain during testing or a battlefront firefight. In recent years a different line of research has put the physiology of stress in an entirely new perspective. The response to stress is not just a primal reaction affecting parts of the brain that are common to a wide array of species ranging from salamanders to humans. Stress, in fact, can cripple our most advanced mental faculties, the areas of the brain most developed in primates.

Older textbooks explained that the hypothalamus, an evolutionarily ancient structure lodged at the base of the brain, reacts to stress by triggering the secretion of a wave of hormones from the pituitary and adrenal glands, which makes the heart race, elevates blood pressure and diminishes appetite. Now research reveals an unexpected role for the prefrontal cortex, the area immediately behind the forehead that serves as the control center that mediates our highest cognitive abilities—among them concentration, planning, decision-making, insight, judgment and the ability to retrieve memories. The prefrontal cortex is the part of the brain that evolved most recently, and it can be exquisitely sensitive to even temporary everyday anxieties and worries.

When things are going well, the prefrontal cortex acts as a control center that keeps our baser emotions and impulses in check. The new research demonstrates that acute, uncontrollable stress sets off a series of chemical events that weaken the influence of the prefrontal cortex while strengthening the dominance of older parts of the brain. In essence, it transfers high-level control over thought and emotion from the prefrontal cortex to the hypothalamus and other earlier-evolved structures. As the older parts take over, we may find ourselves either consumed by paralyzing anxiety or subject to impulses that we usually manage to keep in check: indulgence in excesses of food, drink, drugs or a spending spree at a local specialty store. Quite simply, we lose it.

The growing understanding that acute stress can severely compromise the function of higher "executive" areas in the human brain has drawn the interest of investigators. They are now not just trying to understand what happens in your head when you freeze but also developing behavioral and pharmaceutical interventions to help you keep your composure.

Mind the Jitters

Why we lose it has fascinated scientists for decades. After World War II, investigators analyzed why pilots who were highly skilled in peacetime made simple but fatal mistakes in maneuvering their craft during the heat of battle. What actually happens behind the human skull's frontal bone remained a mystery until the relatively recent arrival of neuroimaging techniques. In a brain scanner, the riot of activity in the prefrontal cortex gives a clue to just how vulnerable the brain's master controller is.

The prefrontal cortex is so sensitive to stress because of its special status within the hierarchy of brain structures. It is the most highly evolved brain region, bigger proportionally in humans than in other primates, and makes up a full third of the human cortex. It matures more slowly than any other brain area and reaches full maturity only after the teen years have passed. The prefrontal area houses the neural circuitry for abstract thought and allows us to concentrate and stay on task, while storing information in the mental sketch pad of working memory. This temporary memory storage area operates by allowing us to keep "in mind" such information as the sum of digits that need to be carried over to the next column when performing addition. As a mental-control unit, the prefrontal area also inhibits inappropriate thoughts and actions.

The neurological executive center functions through an extensive internal network of connections among the triangular-shaped neurons called pyramidal cells. These neurons also send out connections to more distant reaches of the brain that control our emotions, desires and habits. When unstressed, the circuits in this network hum along contentedly. Working memory reminds us to start that assignment due next week, and other circuitry sends a message to lower brain regions signaling that it is perhaps best to forgo a second glass of wine. Meanwhile a message to the amygdala, a deep-brain structure that controls fear reactions, provides assurance that the huge hulk approaching on the sidewalk is not about to smash you in the face.

Keeping this network firing as it should can be a fragile process—and when stress hits, even small changes in the neurochemical environment can instantly weaken network connections. In response to stress, our brain floods with arousal chemicals such as norepinephrine and dopamine, which are released by neurons in the brain stem that send projections throughout the brain. Elevated levels of these signaling chemicals in the prefrontal cortex shut off neuron firing, in part by weakening the connection points, or synapses, between neurons temporarily. Network activity diminishes, as does the ability to regulate behavior. These effects only worsen as the adrenal glands near the kidneys, on command from the hypothalamus, spritz the stress hormone cortisol into the bloodstream, sending it to the brain. In this circumstance, self-control depends on a tricky balancing act.

"Keeping one's cool" is an expression that accurately represents the underlying biological processes. The neural machinery of the prefrontal cortex–and its ability to muster working memory to stay focused on the task at hand–may keep the cascade of neurotransmitters generated deep within the brain from triggering a panicked tide of emotion.

Our research clarifying how easily the prefrontal cortex can be shut down started about 30 years ago. Studies in animals by one of us (Arnsten), along with the late Patricia Goldman-Rakic of Yale University, were among the first to illustrate how neurochemical changes during stress can rapidly switch off prefrontal function. The work showed that neurons in the prefrontal cortex disconnect and stop firing after being exposed to a flood of neurotransmitters or stress hormones.

In contrast, areas deep within the brain take a stronger hold over our behavior. Dopamine arrives at a series of deep-brain structures, collectively called the basal ganglia, that regulate cravings and habitual emotional and motor responses. The basal ganglia hold sway not only when we ride a bicycle without falling but also when we indulge in addictive habits, such as those that make us long for that forbidden ice cream.

In 2001 Benno Roozendaal, now at Radboud University in the Netherlands, James McGaugh of the University of California, Irvine, and their colleagues found similar changes in the amygdala, another older brain region. In the presence of norepinephrine and cortisol, the amygdala alerts the rest of the nervous system to prepare for danger and also strengthens memories that are related to fear and other emotions.

This research now extends to humans. These studies have begun to show that some people seem more vulnerable than others because of their genetic makeup or because of a previous history of stress exposure. After dopamine and norepinephrine switch off circuits in the prefrontal area required for higher cognition, enzymes normally chew up the neurotransmitters so that the shutdown does not persist. In this way, we can return to our baseline when stress abates. Certain forms of a gene can weaken these enzymes, making people more vulnerable to stress and, in some cases, mental illness. Similarly, environmental factors can increase vulnerability; for example, lead poisoning can mimic aspects of the stress response and erode cognition.

Still other research focuses on what happens when the assault on the prefrontal cortex lasts for days or weeks. Chronic stress appears to expand the intricate web of connections among neurons in our lower emotional centers, whereas the areas engaged during flexible, sustained reasoning–anything from the philosophy of Immanuel Kant to calculus–start to shrivel. Under these conditions, the branching, signal-receiving dendrites in the primal amygdala enlarge, and those in the prefrontal cortex shrink. John Morrison of the Icahn School of Medicine at Mount Sinai and his colleagues have shown that prefrontal dendrites can regrow if the stress disappears, but this ability to rebound may vanish if the stress is especially severe. One of us (Sinha) has found evidence of this in humans, where the shrinkage in prefrontal gray matter relates to the history of stress exposure.

This chain of molecular events makes us more vulnerable to subsequent stress and most likely contributes to depression, addiction and anxiety disorders, including post-traumatic stress. Gender appears to be a factor in determining how we react to stress. In women, the hormone estrogen may amplify sensitivity. For example, as one of us (Mazure) and her colleagues have shown, life stress poses a greater risk for depression and is more likely to reduce abstinence from certain addictive behaviors, such as smoking, in women as compared with men. In men, stress may play a more prominent role in exacerbating cravings and eliciting habitual behaviors mediated by the basal ganglia. Reactions that are linked to a specific life stressor are set in a psychosocial context that also can be different for women and men. In ongoing research on frontline health-care workers confronting health risks associated with the COVID-19 pandemic, Mazure and her colleagues are finding that women may have higher rates of stress-related symptoms in response to this experience, and they also have higher rates of concurrent "background stress" such as family or caregiving responsibilities and lower income. This suggests that, for women, biological systems may be chronically on alert, thus presenting an increased challenge in negotiating a significant life stress.

More work on how stress alters the brain's prefrontal self-control locus remains to be done. Some researchers are investigating how other neurochemicals affect the prefrontal cortex. Trevor W. Robbins and Angela Roberts, both at the University of Cambridge, are looking at whether serotonin, which plays a key role in depression, may modulate stress and anxiety through its actions in the prefrontal cortex.

These studies remain challenging because modern ethical standards for experiments using humans state that subjects should not be exposed to situations of extreme psychological stress, and indeed human study participants are told they can stop at any time, giving them control over the experimental situation in a manner that does not mimic real-life stress. Several laboratories have succeeded in simulating the effects of uncontrolled stress by having study participants watch disturbing movies or, as done by the Sinha group, briefly imagine their own stressful experiences to tap into their reactions.

One question that still perplexes researchers is why the brain has built-in mechanisms to weaken its highest cognitive functions. We still do not know for sure, but the triggering of these primal reactions may perhaps have saved human lives when a predatory wild animal was lurking in the bushes. If we suddenly see a tiger burning bright in the forest, it is far more useful to freeze so that the animal cannot see us than to be remembering the words of William Blake's poem.

Absent our slow, deliberate higher-brain networks, primitive brain pathways can stop us on a dime or ready us to flee. These mechanisms may serve a similar function when we face danger in the modern world—say, when a reckless driver cuts us off and we need to slam on the brakes. If we remain in this state, though, prefrontal function weakens, a devastating handicap in circumstances where we need to engage in complex decision-making about a loved one's serious medical condition or organize an important project on a tight deadline.

Get Ahold of Yourself

A logical response to our growing understanding of the jitters is to devise strategies to keep our neural-control center intact. Scientists hope that understanding the molecular events that cause the brain to degenerate from a "reflective" to a "reflexive" state may lead to better treatments for stress disorders. Some of these insights confirm what we already know. Training for emergencies or for military service is all about teaching the basal ganglia and other brain structures to learn the automatic reactions needed to survive. Animal research suggests that the sense of psychological control that becomes second nature to a soldier or emergency medical technician remains the deciding factor in whether we fall apart during stress. Public speaking exhilarates those who feel confident before an audience. For others, it induces nothing but terror, and their minds "go blank."

The routines of the drill sergeant are mirrored by animal studies that show that juveniles grow up to be more capable in handling stress if they have had multiple, successful experiences confronting mild stress in their youth. Similarly, human studies indicate that success in managing challenging situations can build resilience. In contrast, if children stumble through these experiences, they can become more sensitive to and burdened by stress and depression when they grow up.

Clues to new treatments may be slowly emerging from the lab. The drug prazosin, a generic therapy for blood pressure that blocks some of norepinephrine's detrimental actions, appears to decrease both alcohol cravings and levels of consumption. In 2011 Sherry McKee of Yale and her colleagues found that another generic medication for blood pressure, called guanfacine, can inhibit some stress reactions and strengthen prefrontal cortical networks, helping people to resist smoking during stress exposure. (Arnsten and Yale University received royalties from Shire Pharmaceuticals for an extended-release form of guanfacine used for treatment of attention-deficit hyperactivity disorder for children and adolescents but did not receive royalties for the immediate-release form of the drug used in adults in this study.) Further, many labs have shown that behavioral strategies such as relaxation, deep breathing and meditation can reduce the stress response.

And what about that sense of control? Perhaps by learning about how the brain reacts to stress, you may come away with an enhanced sense of control. So maybe the next time you are taking a test or speaking in public and your mind goes blank, you can say to yourself, "This is just my brain trying to save me from a tiger." Maybe it will bring a comforting smile to your face even if it does not bring the correct answer or word to mind.

About the Authors

Amy Arnsten is a professor of neurobiology at the Yale School of Medicine. Her research on molecular changes in the prefrontal cortex during stress and aging has led to treatments such as prazosin and guanfacine for post-traumatic stress disorder, attention-deficit hyperactivity disorder and other conditions.

Carolyn M. Mazure is Spungen Bildner Professor in Women's Health Research and a professor of psychiatry and psychology at the Yale School of Medicine. She created and directs Yale's interdisciplinary research center on the health of women.

Rajita Sinha directs the Yale Stress Center, which focuses on understanding the effects of stress on behavior. She is a professor of psychiatry at the Yale School of Medicine.

"Stress Hormone" Cortisol Linked to Early Toll on Thinking Ability

By Karen Weintraub

The stresses of everyday life may start taking a toll on the brain in relatively early middle age, new research shows. The study of more than 2,000 people, most of them in their 40s, found those with the highest levels of the stress-related hormone cortisol performed worse on tests of memory, organization, visual perception and attention.

Higher cortisol levels, measured in subjects' blood, were also found to be associated with physical changes in the brain that are often seen as precursors to Alzheimer's disease and other forms of dementia, according to the study published in October in *Neurology*.

The link between high cortisol levels and low performance was particularly strong for women, the study found. But it remains unclear whether women in midlife are under more stress than men or simply more likely to have their stress manifested in higher cortisol levels, says lead researcher Sudha Seshadri. A professor of neurology, she splits her time between Boston University and The University of Texas Health Science Center at San Antonio, where she is the founding director of the Glenn Biggs Institute for Alzheimer's & Neurodegenerative Diseases.

Working on the study "made me more stressed about not being less stressed," Seshadri says, laughing. But, she adds, the bottom line is serious: "An important message to myself and others is that when challenges come our way, getting frustrated is very counterproductive–not just to achieving our aims but perhaps to our capacity to be productive."

The study is the largest of its kind to look at these factors and tightens the link between cortisol, midlife stress and brain changes, says Pierre Fayad, medical director of the Nebraska Stroke Center at the University of Nebraska Medical Center, who was not involved in the new research. "It confirms some of the previous suspicions," he says. "Because of its quality, it gives a lot more credibility."

Bruce McEwen, a neuroscientist and cortisol expert at The Rockefeller University who also was not part of the study, says he found it "frankly remarkable." Cortisol, he notes, is necessary for life–so it is obviously not all bad. But stress can lead people to potentially problematic behaviors such as smoking, drinking and eating unhealthy food. "Cortisol is itself the tip of the iceberg of things that are going on in a person's life and a person's body," he says.

The new research included volunteers from the Framingham Heart Study, a 70-year-old study of residents from a Boston suburb. Researchers are now studying the grandchildren of the original participants, most of whom were white, middle class and suburban, Seshadri says. Although the scientists did not ask participants what kinds of specific stresses they were under on the day their blood was drawn, she says the volunteers were able to come in for a three-to-four-hour examination–so "you would say they were at a reasonably stable point in their life."

Yet even these relatively young and apparently well-off people showed signs of brain changes, both in brain scans and in their performance. "This is the range of stress that a group of average Americans would experience," Seshadri says. The highest cortisol levels were associated with changes that could be seen on an MRI scan of the brain, the study found.

Cortisol does not distinguish between physical and mental stress, so some of the people with high levels might have had physical illnesses such as diabetes that drove up their cortisol levels, Seshadri says. It is also possible levels of the hormone might spike in people's blood if they are already undergoing brain changes–that is, the elevated cortisol could be the result of the changes rather than their cause–she says. But she thinks this is unlikely because the trial participants were so young. Each subject's cortisol level was measured only once (in the morning), so the measurements do not reflect changes over time or variations throughout the day, she notes.

The volunteers were given tasks such as copying a shape they were shown, or being asked to repeat a story they had been told 20 minutes earlier. The differences in performance were subtle, Seshadri says. She could not immediately tell whether subjects had higher or lower cortisol levels based on how well they carried out the tasks. "It was more that in terms of group averages there was a real difference," she explains.

Earlier research has shown weaker-than-average performances on tests like these are associated with a higher risk of dementia decades later, and Seshadri says high stress levels in midlife might be one of many factors that contribute to dementia. Understanding that link might offer a potential opportunity to reduce risk–but she cautions research has not yet shown conclusively that lowering cortisol levels will reduce the risk of Alzheimer's.

Other research has shown cortisol levels can be reduced with adequate sleep, exercise, socializing and relaxing mental activities such as meditation. "There are a number of intriguing, fairly simple things that have been shown to change these levels," Seshadri says. "But whether they will in turn translate into better preservation of the brain is something that can only be determined in a clinical trial."

Rockefeller University's McEwen says other research suggests it is never too late to adopt a healthier lifestyle by taking steps like reducing stress, exercising regularly, eating a healthy diet, getting enough good-quality sleep and finding meaning in one's life. "The life course is a one-way street," he says. But "the brain does have the capacity for repairing."

About the Author

Karen Weintraub is a staff writer at USA Today, *where she covers COVID, vaccine development and other health issues.*

Brain's Dumped DNA May Lead to Stress, Depression

By Knvul Sheikh

Humans and other mammals react to stressful situations through a series of well-orchestrated evolutionary adaptations. When faced with a predator looking for its next meal, or with worry of losing a job, our bodies release a cascade of stress hormones. Our heart rate spikes, breath quickens, muscles tense up and beads of sweat appear.

This so-called "fight-or-flight" response served our ancestors well, but its continual activation in our modern-day lives comes with a cost. Scientists are starting to realize stress often exacerbates several diseases, including depression, diabetes, cardiovascular disease, HIV/AIDS and asthma. One theory is hoping to explain the link between stress and such widespread havoc by laying the blame on an unexpected source–the microscopic powerhouses inside each cell.

Each of our cells contains hundreds of small bean-shaped mitochondria – subcellular structures, or organelles, that provide the energy needed for normal functioning. Mitochondria have their own circular genome with 37 genes. We inherit this mitochondrial DNA only from our mothers, so the makeup of the DNA's code stays relatively consistent from one generation to the next.

But our fight-or-flight response places extreme demands on the mitochondria. All of a sudden, they need to produce much more energy to fuel a faster heartbeat, expanding lungs and tensing muscles, which leaves them vulnerable to damage. Unlike DNA in the cell's nucleus, though, mitochondria have limited repair mechanisms. And recent animal studies have shown chronic stress not only leads to mitochondrial damage in brain regions such as the hippocampus, hypothalamus and cortex, it also results in mitochondria releasing their DNA into the cell cytoplasm, and eventually into the blood.

The genetic cast-offs are not just inert cellular waste. "This circulating mitochondrial DNA acts like a hormone," says Martin Picard, a psychobiologist at Columbia University, who has been studying mitochondrial behavior and the cell-free mitochondrial DNA for the better part of the last decade. Ejection of mitochondrial DNA from the cell mimics somewhat adrenal glands' release of cortisol in response to stress, he says. Certain cells produce the circulating mitochondrial DNA and, as with the adrenal glands, its release is also triggered by stress.

To demonstrate psychological stress can cause mitochondrial DNA to be released by cells, Picard and his team devised a quick stress test. They asked 50 otherwise healthy men and women to deliver a quick speech defending themselves against a false accusation on camera. Afterward the researchers took blood samples from the participants and compared them with blood taken immediately before they were stressed. Even though the stressful task only lasted a total of five minutes, the scientists found participants' serum circulating mitochondrial DNA levels more than doubled 30 minutes after the test. These results, currently under review, provide the first direct evidence for how bits of mitochondrial DNA floating in our blood may relay stress to other parts of the body, like dominoes tumbling one after another.

Previous studies have provided several clues that suggest circulating mitochondrial DNA is a hallmark of stress. In 2016 Swedish researchers published findings in *Translational Psychiatry* demonstrating elevated levels of mitochondrial DNA outside the cell in 37 people who had recently attempted suicide. Earlier this year the same group of scientists published another paper in *Neuropsychopharmacology* showing people with major depression had high levels of circulating mitochondrial DNA, and these levels kept increasing in patients who did not respond well to antidepressant medication.

These studies are all part of an emerging field of research on mitochondrial DNA, where scientists are recognizing that the tiny organelles have effects across a wide range of diseases.

"Mitochondrial DNA is probably the most sensitive thing in your body," says Douglas Wallace, director of the Center for Mitochondrial and Epigenomic Medicine at The Children's Hospital of Philadelphia. "If your mitochondria are sensing a problem, then all the rest of you is in trouble, too."

In his own research Wallace has shown mitochondrial DNA mutations are more common in people with autism spectrum disorders than in neurotypical adults. Other studies in the past few years have linked mitochondrial dysfunction to schizophrenia, Alzheimer's, arthritis and cancer–all problems where inflammation is also known to occur, Picard notes.

But how was this inflammation triggered by mitochondrial DNA leaking out of cells? A 2010 *Nature* paper provided the answer: In it researchers demonstrated the way mitochondrial DNA, when released into the blood after an injury, mobilized a pro-inflammatory immune response. Because of mitochondria's bacterial origin and its circular DNA structure, immune cells think it's a foreign invader. When circulating mitochondrial DNA binds to a particular receptor, TLR9, on immune cells, they respond as if they were reacting to a foreign invader such as a flu virus or an infected wound. The immune cells release chemicals called cytokines telling other white blood cells they need to report for duty at sites of infection, inflammation or trauma.

Together, this growing understanding of circulating mitochondrial DNA sets a time frame for how psychological stress may lead to widespread inflammation, Picard says. "Mitochondria are the missing link between our psychological state and neurological or other disorders involving inflammation," he says.

It is an interesting shift away from the traditional, anatomical aspects of disease, such as brain shrinkage in depressed patients. But reducing disorders like depression to brain imbalances or shrinkage simply has not explained everything, says Bruce McEwen, a neuroendocrinologist at The Rockefeller University. "If that was the case, you could take Prozac or [selective] serotonin reuptake inhibitors [SSRIs] to fix it, but everybody is now realizing that that's not the way it works," he says. "Otherwise, antidepressants would be more effective."

If further evidence of the importance of healthy mitochondria continues to emerge, drugs that focus on regulating cellular energy production instead could become a new line of defense for psychiatric and biological disorders.

Childhood Stress Decreases Size of Brain Regions

By Christie Nicholson

Children who experience neglect, abuse and poverty have a tougher time as adults than do well-cared-for kids. Now there's evidence that such stress can actually change the size of brain structures responsible for learning, memory and processing emotion. The finding is in the journal *Biological Psychiatry*.

Researchers took images of the brains of 12-year-olds who had suffered either physical abuse or neglect or had grown up poor. From the images the scientists were able to measure the size of the amygdala and hippocampus–two structures involved in emotional processing and memory. And they compared the sizes of these structures with those of 12-year-old children who were raised in middle-class families and had not been abused. And they found that the stressed children had significantly smaller amygdalas and hippocampuses than did the kids from the more nurturing environments.

Early stress has been associated with depression, anxiety, cancer and lack of career success later on in adulthood. This study on the sizes of brain regions may offer physiological clues to why what happens to toddlers can have such a profound impact decades later.

–Christie Nicholson

[The above text is a transcript of a podcast.]

Why Do We Crave Sweets When We're Stressed?

By Achim Peters

Although our brain accounts for just 2 percent of our body weight, the organ consumes half of our daily carbohydrate requirements–and glucose is its most important fuel. Under acute stress the brain requires some 12 percent more energy, leading many to reach for sugary snacks.

Carbohydrates provide the body with the quickest source of energy. In fact, in cognitive tests subjects who were stressed performed poorly prior to eating. Their performance, however, went back to normal after consuming food.

When we are hungry, a whole network of brain regions activates. At the center are the ventromedial hypothalamus (VMH) and the lateral hypothalamus. These two regions in the upper brain stem are involved in regulating metabolism, feeding behavior and digestive functions. There is, however, an upstream gatekeeper, the nucleus arcuatus (ARH) in the hypothalamus. If it registers that the brain itself lacks glucose, this gatekeeper blocks information from the rest of the body. That's why we resort to carbohydrates as soon as the brain indicates a need for energy, even if the rest of the body is well supplied.

To further understand the relationship between the brain and carbohydrates, we examined 40 subjects over two sessions. In one, we asked study participants to give a 10-minute speech in front of strangers. In the other session they were not required to give a speech. At the end of each session, we measured the concentrations of stress hormones cortisol and adrenaline in participants' blood. We also provided them with a food buffet for an hour. When the participants gave a speech before the buffet, they were more stressed, and on average consumed an additional 34 grams of carbohydrates, than when they did not give a speech.

So what about that chocolate, then? If a person craves chocolate in the afternoon, I advise him or her to eat chocolate to stay fit and keep his or her spirits up. That's because at work people are often stressed and the brain has an increased need for energy. If one doesn't eat anything, it's possible the brain will use glucose from the body, intended for fat and muscle cell use, and in turn secrete more stress hormones. Not only does this make one miserable, it can also increase the risk of heart attacks, stroke or depression in the long run. Alternatively, the brain can save on other functions, but that reduces concentration and performance.

In order to meet the increased needs of the brain, one can either eat more of everything, as the stressed subjects did in our experiment, or make it easy for the body and just consume sweet foods. Even babies have a pronounced preference for sweets. Because their brain is extremely large compared with their tiny bodies, babies require a lot of energy. They get that energy via breast milk, which contains a lot of sugar. Over time, our preference for sweets decreases but never completely disappears, even as we become adults. The extent to which that preference is preserved varies from person to person and seems to depend, among other things, on living conditions. Studies suggest people who experience a lot of stress in childhood have a stronger preference for sweets later in life.

For some, the brain cannot get its energy from the body's reserves, even if there are enough fat deposits. The most important cause of this is chronic stress. To ensure their brains are not undersupplied, these people must always eat enough. Often the only way out of such eating habits is to leave a permanently stressful environment. So although many tend to be hard on themselves for eating too many sweets or carbs, the reasons behind such craving aren't always due to a lack of self-control and might require a deeper look into lifestyle and stressful situations—past and present. Once the root cause of stress is addressed, eating habits could ultimately resolve themselves.

About the Author

Achim Peters is a brain researcher and diabetologist. He leads the Selfish Brain clinical research group at the University of Lübeck and has authored two books on how the selfish brain influences weight under chronic stress.

Strain on the Brain

By Brian Mossop

A recent wave of research has unveiled an important environmental player in the genesis of neurodegenerative disease: stress.

Pairs of identical twins developed Alzheimer's disease in concert only 40 percent of the time, showing that factors other than genetics must contribute to the disorder.

Stress seems to impede the ability of certain brain cells to recover from insults, triggering or aggravating the symptoms of disorders such as Parkinson's.

In 2007 Nobel laureate James Watson eyed his genome for the very first time. Through more than 50 years of scientific and technological advancement, Watson saw the chemical structure he once helped to unravel now pieced into a personal genetic landscape that lay before him.

There was one small stretch of DNA on chromosome 19, however, that he chose to leave under wraps. That region coded for the *apolipoprotein E* (*APOE*) gene. Since the early 1990s *APOE* has been a telling genetic marker of Alzheimer's risk: certain forms of it correlate strongly with the development of the disease. Watson's grandmother suffered from Alzheimer's, but without any reasonable treatments or proved preventive strategies, the discoverer of the double helix decided the information was too volatile, its revelation creating more potential harm than good.

Watson's apprehension is understandable. Treatments for Alzheimer's have consistently failed. But as scientists learn more and more about the brain, they have come to realize that genetics alone rarely dictates the course of an illness. Instead brain disorders result from a complex interaction between our genes and the environments to which we are exposed. Indeed, a set of recent studies has just uncovered an important environmental instigator of neurodegenerative disease: stress.

Researchers have catalogued the effect of stress on numerous psychological conditions, including depression and chronic anxiety. The idea that stress may figure into neurodegenerative diseases, however, is relatively new. Although the notion that our high-pressure jobs and hectic lives might be doing additional damage could be worrisome, stress is at least something we can theoretically control. That is, trying to relax might be a first step toward raising the chances of keeping your brain free of disease in old age.

Tight Quarters

Since Alois Alzheimer first documented "presenile dementia" in a patient at the beginning of the 20th century, doctors have often observed that the disease runs in families. But not until the early 1990s, about the same time the *APOE* link surfaced, did researchers glean hints that nongenetic factors contribute to it.

Epidemiologist Brenda Plassman of Duke University and her colleagues teased out this environmental effect by studying identical twins, who share virtually the same genetic material. If a disease is driven purely by genetics, then when one twin develops it, the other will be stricken as well. By analyzing data from a large cohort of identical twins (all of them male veterans of World War II) collected by the National Academy of Sciences and the National Research Council, Plassman and her co-workers reported in 2000 that when one twin developed Alzheimer's, the other twin developed the disease only 40 percent of the time. Concluding that factors besides genetics must be at play, the investigators have since been searching for those contributors. Among the possibilities: subtle medical conditions, occupational characteristics and physical activity levels.

Of course, your job and the amount you exercise both have an effect on your level of psychological stress, the mind and body's response to challenge and change. But only this year did evidence suggest that stress might be a key ingredient in the recipe for cognitive decline. To explore how different environments might affect the development of Alzheimer's, neuroscientist Mark H. Tuszynski of the University of California, San Diego, and his colleagues examined the brains of aged rhesus monkeys that had spent their early lives in either small or standard-size cages. Tight quarters have been shown to stress these animals, elevating levels of glucocorticoid hormones in their blood. The exact cause of this hormonal rise—whether it comes from a feeling of being trapped or an inability to get adequate exercise, or both—is still an open question.

Cortisol, a glucocorticoid hormone released when humans experience stress, influences the brain through specialized molecular receptors on neurons in a number of brain regions. When cortisol binds to its receptor, the interaction triggers molecular events that reduce communication at synapses, the junctions between neurons, which may ultimately cause the connections to wither away. Using protein stains that adhere specifically to synapses, enabling them to be seen, Tuszynski's team determined the relative number of synapses in all the monkeys. Using a similar method, the researchers also assessed the amount of sticky amyloid plaques, a pathological hallmark of Alzheimer's.

Compared with the monkeys raised in standard-size cages, those that lived in smaller cages had, on average, a significantly higher density of plaques and fewer synapses in one part of their brain—the same pattern seen in the brains of Alzheimer's patients at autopsy. The finding suggests that the size of an animal's cage—and perhaps the amount of stress it endures as a result—may shape that animal's brain in a way that affects its vulnerability to certain types of degeneration as it ages. Interestingly, the amount of plaque riddling the brains of the monkeys housed in smaller cages varied a lot, indicating that stress affects individuals differently. After all, we all know people who seem to take even mildly negative events to heart as well as others in similar situations who take their plight in stride.

The evidence from Tuszynski's group has its limitations. Observations in monkeys living in labs do not precisely mirror the human condition. In addition, these findings correlate only one aspect of early-life experience with pathological signs of degeneration. We do not know that the stress caused the changes, nor do we know whether those changes resulted in true cognitive slipups, because the scientists could not test the animals' cognitive function.

Toxic Tension

Nevertheless, additional studies in rodents suggest that even intermittent strain can tip the scales toward dementia, even if it does not lead to cognitive breaks on its own. In March 2010 neuropharmacologist Karim Alkadhi of the University of Houston and his colleagues put rats at risk for dementia by injecting them with very low concentrations of beta-amyloid peptides, the molecules that form plaques in humans. The researchers then stressed some of the animals by placing an intruder rat in their home cage. As expected, blood levels of corticosterone, a glucocorticoid, rose in the stressed rats.

Then the scientists placed each rat in a water tank containing a maze. A rat had to find the path that led to a platform to escape the water—a rodent test of learning and memory. Usually after a few tries, a rat will remember the correct route; it will then swim directly to the platform, even a day or two later. Most of the experimental rats—including those that had been given amyloid injections and those forced to face intruders—performed well. The rats that had received both the shots and the unwanted visitor, however, had difficulty. So although stress alone does not degrade memory, it does seem to push at-risk animals over the edge, making them less able to learn and remember new things.

Other work hints that stress may hasten the onset of Parkinson's disease, a neurodegenerative disorder characterized by motor difficulties rather than cognitive deficits. The loss of brain cells that produce dopamine, a neurotransmitter essential for voluntary movement, causes Parkinson's patients to shake, become rigid and lose coordination.

To re-create these deficits in rats, behavioral neuroscientist Gerlinde A. S. Metz and her colleagues at the University of Lethbridge in Alberta infused a toxic drug into a brain area rich with dopamine neurons. Some of these animals were put into a Plexiglas tube for 20 minutes a day for two weeks, producing a temporary boost in stress hormone levels. Another group received corticosterone shots, which kept the animals' stress hormones high throughout the experiment. Metz's team then tested the motor skills of all the animals. In one exercise, for example, the rats had to slip their paws through a narrow opening in a Plexiglas box to extract a small food pellet, an action that requires precise and careful movements.

The Metz team's toxic treatment is transient; usually the treated rats' motor skills improve with time. But the animals with elevated corticosterone levels–both the ones that spent time in a stressful environment and those that received hormone shots–continued to struggle with the pellet extraction task long after the other animals had recovered. The results suggest that stress impedes the ability of dopamine cells to recover from insults, triggering or aggravating Parkinson's symptoms.

Indelible Mark

Using such eye-opening studies as these, scientists are learning that stress is more than a fleeting emotional setback. Rather, in certain situations, stress can leave an indelible mark on our brain.

But there is good news, too. Stress is a contributor to neurodegeneration that can be controlled. Just as many individuals with high cholesterol levels now take preemptive action to stave off heart disease, one day people may use, say, their APOE status to motivate them to adjust their lifestyles. Evidence suggests that simple interventions such as exercise, meditation and getting enough sleep can help reduce the stress of life's encounters. Such measures might even ease the anxiety of knowing which APOE stamp adorns your genome.

Referenced

Acute Stress Modulates Genotype Effects on Amygdala Processing in Humans. Helena Cousijn et al. in *Proceedings of the National Academy of Sciences USA*, Vol. 107, No. 21, pages 9867–9872; May 25, 2010. www.pnas.org/content/107/21/9867.full

Association of Early Experience with Neurodegeneration in Aged Primates. David A. Merrill et al. in *Neurobiology of Aging*, Vol. 32, No. 1, pages 151–156; January 2011.

Metz Behavioral Neuroscience Lab: http://people.uleth.ca/~gerlinde.metz/home.htm

About the Author

Brian Mossop holds a Ph.D. in biomedical engineering and has had postdoctoral training in neuroscience. He also writes for Scientific American, Wired, Slate *and* The Scientist *and is the community manager at the Public Library of Science (PLoS).*

Section 3: Causes of Stress

How Parents' Trauma Leaves Biological Traces in Children

By Rachel Yehuda

After the twin towers of the World Trade Center collapsed on September 11, 2001, in a haze of horror and smoke, clinicians at the Icahn School of Medicine at Mount Sinai in Manhattan offered to check anyone who'd been in the area for exposure to toxins. Among those who came in for evaluation were 187 pregnant women. Many were in shock, and a colleague asked if I could help diagnose and monitor them. They were at risk of developing post-traumatic stress disorder, or PTSD—experiencing flashbacks, nightmares, emotional numbness or other psychiatric symptoms for years afterward. And were the fetuses at risk?

My trauma research team quickly trained health professionals to evaluate and, if needed, treat the women. We monitored them through their pregnancies and beyond. When the babies were born, they were smaller than usual—the first sign that the trauma of the World Trade Center attack had reached the womb. Nine months later we examined 38 women and their infants when they came in for a wellness visit. Psychological evaluations revealed that many of the mothers had developed PTSD. And those with PTSD had unusually low levels of the stress-related hormone cortisol, a feature that researchers were coming to associate with the disorder.

Surprisingly and disturbingly, the saliva of the nine-month-old babies of the women with PTSD also showed low cortisol. The effect was most prominent in babies whose mothers had been in their third trimester on that fateful day. Just a year earlier a team I led had reported low cortisol levels in adult children of Holocaust survivors, but we'd assumed that it had something to do with being raised by parents who were suffering from the long-term emotional consequences of severe trauma. Now it looked like trauma could leave a trace in offspring even before they are born.

In the decades since, research by my group and others has confirmed that adverse experiences may influence the next generation through multiple pathways. The most apparent route runs through parental behavior, but influences during gestation and even changes in eggs and sperm may also play a role. And all these channels seem to involve epigenetics: alterations in the way that genes function. Epigenetics potentially explains why effects of trauma may endure long after the immediate threat is gone, and it is also implicated in the diverse pathways by which trauma is transmitted to future generations.

The implications of these findings may seem dire, suggesting that parental trauma predisposes offspring to be vulnerable to mental health conditions. But there is some evidence that the epigenetic response may serve as an adaptation that might help the children of traumatized parents cope with similar adversities. Or could both outcomes be possible?

In the Aftermath

My first encounter with intergenerational transmission of trauma was in the 1990s, soon after my team documented high rates of PTSD among Holocaust survivors in my childhood community in Cleveland. The first study of its kind, it garnered a lot of publicity; within weeks I found myself heading a newly created Holocaust research center at Mount Sinai staffed largely by professional volunteers. The phone was ringing off the hook. The callers weren't all Holocaust survivors, though; most were the adult children of Holocaust survivors. One particularly persistent caller—I'll call him Joseph—insisted that I study people like him. "I'm a casualty of the Holocaust," he claimed.

When he came in for an interview, Joseph didn't look like a casualty of anything. A handsome and wealthy investment banker in an Armani suit, he could've stepped off the pages of a magazine. But Joseph lived each day with a vague sense that something terrible was going to happen and that he might need to flee or fight for his life.

He'd been preparing for the worst since his early 20s, keeping cash and jewelry at hand and becoming proficient in boxing and martial arts. Lately he was tormented by panic attacks and nightmares of persecution, possibly triggered by reports of ethnic cleansing in Bosnia.

Joseph's parents had met in a displaced-persons camp after surviving several years at Auschwitz, then arrived penniless in the U.S. His father worked 14 hours a day and said very little, never mentioning the war. But almost every night he woke the family with shrieks of terror from his nightmares. His mother spoke endlessly about the war, telling vivid bedtime stories about how relatives had been murdered before her eyes. She was determined that her son succeed, and his decision to remain unattached and childless infuriated her. "I didn't survive Auschwitz so that my own child would end the family line," she'd say. "You have an obligation to me and to history."

We ended up talking to many people like Joseph: adult children of Holocaust survivors who suffered from anxiety, grief, guilt, dysfunctional relationships and intrusions of Holocaust-related imagery. Joseph was right–I needed to study people like him. Because those who were calling us were (in research-speak) self-selecting, we decided to evaluate the offspring of the Holocaust survivors we had just studied in Cleveland. The results were clear. Survivors' adult children were more likely than others to have mood and anxiety disorders, as well as PTSD. Further, many Holocaust offspring also had low cortisol levels–something that we had observed in their parents with PTSD.

Fight, Flight—or Freeze

What did it all mean? Unraveling the tangle of trauma, cortisol and PTSD has occupied me and many other researchers for the decades since. In the classic fight-or-flight response, identified in the 1920s, a threatening encounter triggers the release of stress hormones such as adrenaline and cortisol. The hormones prompt a cascade of changes, such as quickening the pulse and sharpening the senses to enable the threatened person or animal to focus on and react to the immediate danger. These acute effects were believed to dissipate once the danger receded.

In 1980, however, psychiatrists and other advocates for Vietnam War veterans won a prolonged struggle to get post-traumatic stress included in the third edition of the *Diagnostic and Statistical Manual of Mental Disorders* (*DSM-III*). It was the first official recognition that trauma could have long-lasting effects. But the diagnosis was controversial. Many psychologists believed that its inclusion in *DSM-III* had been politically, rather than scientifically, driven—in part because there were no scientific explanations for how a threat could continue to influence the body long after it was removed.

Complicating matters, studies of Vietnam veterans were generating perplexing results. In the mid-1980s neuroscientists John Mason, Earl Giller and Thomas Kosten, then all at Yale University, reported that veterans with PTSD had high levels of adrenaline but lower levels of cortisol than patients with other psychiatric diagnoses. Because stress usually causes stress hormones, including cortisol, to rise, many researchers, including myself, were skeptical of these observations. When I joined the Yale laboratory as a postdoctoral fellow a year later, I studied a different group of veterans using other methods for measuring cortisol. To my astonishment, I replicated the finding.

I still couldn't believe that the low cortisol levels had anything to do with trauma. Surely the Holocaust was as terrible as the Vietnam War, I reasoned—and growing up as a rabbi's daughter in a community full of Holocaust survivors, many of them my friends' parents, I'd noticed nothing out of the ordinary about them. I was sure that they didn't have either PTSD or low cortisol, I told my mentor, Giller. "That's a testable hypothesis," he responded. "Why don't you study that, instead of conjecturing?"

So my team of five people landed in Cleveland, along with a centrifuge and other equipment. We stayed in my parents' home, walking door to door to interview people by day and returning to test blood and urine samples in the evening. When the results came in, they were clear: half the Holocaust survivors had PTSD, and those with PTSD had low cortisol. There was no question about it—even if the traumatic experience happened long ago, PTSD went hand in hand with low cortisol.

But why? And which came first? An important clue came from a 1984 review by the late Allan Munck and other researchers at the Geisel School of Medicine at Dartmouth. They noted that among stress hormones, cortisol played a special, regulatory role. High levels of stress hormones, if sustained for a long time, harm the body in multiple ways, weakening the immune system and increasing susceptibility to problems such as hypertension. But in a context of acute trauma, cortisol may paradoxically also have a protective effect. It shuts down the release of stress hormones—including itself—reducing the potential damage to organs and the brain. Such a trauma-induced feedback loop could conceivably reset the cortisol "thermostat" to a lower level.

I picked up another piece of the puzzle and placed it. In the early 1990s we'd shown that Vietnam veterans were more likely to develop PTSD if they'd been abused as children. Slowly a shape was emerging that connected intense childhood adversity—a period of "freeze" because a child usually cannot fight or flee—with low cortisol and the possibility of future PTSD. We studied people who'd been raped or who'd been in auto accidents when they came into emergency rooms, finding that those with lower cortisol levels were more likely to develop PTSD after the attack or accident.

Could low cortisol levels have been present before the event that brought them into the emergency room? I wondered. If someone with low cortisol was subjected to a traumatic experience, we reasoned, the cortisol levels in their bodies might be too low to tamp down the stress reaction. Adrenaline levels might then shoot way up, searing the memory of the new trauma into the brain—from where it might later surface as flashbacks or nightmares. Perhaps low cortisol marked a vulnerability to developing PTSD.

The study of Holocaust offspring supported this conjecture. Children of Holocaust survivors with PTSD tended to have low cortisol even if they did not have their own PTSD. As we'd suspected, low cortisol seemed related to vulnerability to PTSD.

The Feedback Loop

But what mechanism connected trauma exposure to low cortisol to future PTSD? We began a series of studies to answer this question. Significantly, we found that Vietnam veterans with PTSD had a greater number of glucocorticoid receptors. These are proteins to which cortisol binds to exert its diverse influences. That suggested a greater sensitivity to cortisol: a small increase in the hormone's concentration would precipitate a disproportionate physiological reaction. But it wasn't until we looked more closely at the molecular underpinnings of cortisol functioning–in part by examining epigenetics–that we understood how exposure to trauma might reset the cortisol feedback loop.

In the 1990s scientists were realizing that the output of our genes is sensitive to factors not written directly into our genetic code. Genes provide the templates for producing proteins. But much like cakes baked using the same ingredients may turn out differently depending on variations in the oven's temperature, how much of those proteins gets produced, or "expressed," depends on the environment. The discovery gave rise to epigenetics, the study of what influences gene expression and how. It proved crucial to understanding both the neurobiology of PTSD and the intergenerational effects of trauma.

Epigeneticists explore the switches that turn gene expression on and off. One such mechanism, called methylation, involves a methyl group–a methane molecule that is missing one of its four hydrogen atoms, leaving a chemical bond free to attach to another atom or molecule. Methylation is a process by which, in the presence of specific enzymes, methyl groups attach to key sites on a strand of DNA or within the complex of DNA and proteins known as chromatin. By occupying these sites like roadblocks on a highway, methyl groups can alter transcription, a basic step in gene expression where a piece of RNA is made from a DNA template. Increased methylation generally impedes RNA transcription, whereas less methylation enhances transcription. These changes are enduring in that they survive normal cell division and require specific enzymes for their removal.

In 2015 our group became one of the first to pinpoint epigenetic changes on stress-related genes of veterans with PTSD. These alterations partially explained why trauma's effects were so persistent, lasting for decades. Specifically we observed reduced methylation in an important region of *NR3C1*, a gene that encodes the glucocorticoid receptor, likely increasing the sensitivity of these receptors.

This epigenetic modification suggests a potential explanation for how trauma might reset cortisol levels. The body regulates the stress response through a complicated feedback mechanism. A rise in cortisol levels will prompt the body to produce less of the hormone, which may drive up the numbers and responsiveness of glucocorticoid receptors. Given the epigenetic and other changes occurring with sustained responses to trauma, the feedback loop might become recalibrated. In people who have already endured trauma, their stress systems might be sensitized and their cortisol levels diminished–increasing their adrenaline response to further trauma and leading to PTSD.

Epigenetic Inheritance

Could some of these epigenetic changes in trauma survivors also be found in the children of trauma survivors? Finding low cortisol in the 9/11 babies back in 2002 had told us that we'd been thinking about some things all wrong. We'd assumed all along that trauma was behaviorally transmitted: Joseph's problems seemed to result from the stressful, bereaved atmosphere in his childhood home. But now it looked like the uterine environment also played a role. So did the sex of the traumatized parent.

In our early studies of Holocaust offspring, we had selected only those people with two parents who were Holocaust survivors. We redid the studies to figure out if the sex of the parent mattered—and it did. Those whose mother (or both parents) had PTSD tended to exhibit lower cortisol levels and showed evidence of more sensitive glucocorticoid receptors. In contrast, those whose fathers, but not mothers, had PTSD showed the opposite effect.

Taking a closer look, we again discovered lower methylation within the glucocorticoid receptor gene, *NR3C1*, in Holocaust offspring whose mothers, or both parents, had PTSD. These changes mirrored what we'd observed in the maternal survivors themselves. But in offspring with only paternal PTSD, we observed more methylation—the opposite effect. These findings raised the possibility that PTSD in mothers and fathers might lead to different epigenetic changes on the glucocorticoid receptor in children.

In a second series of studies beginning in 2016, we examined methylation within another gene, *FKBP5*, which encodes a protein involved in regulating the ability of the glucocorticoid receptor to bind cortisol. The findings showed related methylation patterns within the *FKBP5* gene in both Holocaust parents and their children. But because of the small number of participants in that study—by this time it was difficult to find many living Holocaust survivors who could participate with their offspring—we couldn't examine how factors such as the parents' PTSD status might contribute to *FKBP5* methylation.

We were able to replicate and extend this work in a substantially larger sample of just the Holocaust offspring, however. In 2020 we reported lower levels of *FKBP5* methylation in the adult children whose mothers—and not fathers—were exposed to the Holocaust during childhood. This effect was independent of whether the mother had PTSD. It suggested that trauma might have affected the mothers' eggs decades before her children were conceived, while she was herself a child.

Given the obvious difficulties in studying generations of people, scientists often resort to animal studies to explore epigenetic transmission. In 2014 Brian Dias and Kerry Ressler, then both at the Emory University School of Medicine, reported an intergenerational epigenetic pathway that ran through sperm. They gave a male mouse a mild electric shock as it smelled a cherry blossom scent, stimulating a fear response to the odor. The response was accompanied by epigenetic changes in its brain and sperm. Intriguingly, the male offspring of the shocked mice demonstrated a similar fear of cherry blossoms—as well as epigenetic changes in their brain and sperm—without being exposed to the shock. These effects were passed down for two generations. In other words, the lesson the grandfather mouse learned, that the cherry blossom scent means danger, was transmitted to its son and grandson.

In a study published in 2020, my colleagues and I experimented with genome-wide gene expression, a tool that can identify links between protein expression and specific conditions across the entire human genome. With this approach, we again observed distinct patterns of gene expression linked with maternal and paternal trauma exposure and PTSD.

In the Womb

Apart from altering the eggs and sperm that encapsulate our genetic inheritance, sometimes decades before conception, trauma also seems to influence the uterine environment. Meticulous studies of the offspring of women who were pregnant during the Dutch Famine—a six-month period during World War II when the Nazis blocked the food supply to the Netherlands, causing widespread starvation—provided an early indication of in utero effects. Researchers discovered that the combined effects of extreme stress and nutritional deprivation, such as deficits in metabolism and susceptibility to cardiovascular illness, depended on the trimester of exposure.

The 9/11 babies were also impacted in the womb, with those in the third trimester having significantly lower cortisol levels. What this condition meant for their future development I sadly never found out. At the wellness visit, mothers who had PTSD (and low cortisol) were more likely to report that their nine-month-olds were unusually anxious and afraid of strangers. But we didn't get the funding to follow the babies into adulthood.

How might the uterine environment leave a trauma trace in the offspring? Our work on Holocaust survivors and their adult children provided some clues. The story is again complicated, and it involves an enzyme known as 11-beta-hydroxysteroid dehydrogenase type 2 (11β-HSD2). Holocaust survivors had lower levels of the enzyme than those who hadn't lived through the Holocaust–and such effects were particularly pronounced in those who were the youngest during World War II. The enzyme is normally concentrated in the liver, kidneys and brain. Under conditions of food deprivation, however, the body can lower levels of 11β-HSD2 to increase metabolic fuel in the interest of promoting survival. In adults, the enzyme level will return to what it had been originally when there is no more starvation, but in children, the level may remain low. Our findings suggested that 11β-HSD2 levels might have been altered during childhood when Holocaust survivors were exposed to long periods of malnourishment; the change persisted well into old age.

In the children of women who were Holocaust survivors, however, we saw quite the opposite: 11β-HSD2 levels were higher than in Jewish control subjects. The result might seem contradictory, but there is a logic to it. During pregnancy, 11β-HSD2 also acts in the placenta, protecting the fetus from exposure to circulating maternal cortisol, which can be toxic to the developing brain. The enzyme, which is particularly active in the third trimester, converts maternal cortisol into an inactive form, creating a kind of chemical shield in the placenta that protects the fetus from the hormone's harmful effects. The high levels of this enzyme in the offspring of Holocaust survivors may thus reflect an adaptation, an effort to protect the fetus from the lowered 11β-HSD2 levels in their mothers.

All of this means that offspring are not always passive recipients of their parents' scars. Just as a parent was able to survive trauma by means of biological adaptations, offspring can sometimes adapt to the biological impact of their parents' trauma.

How traumatized parents interact with their children, of course, also influences their development. One of the most powerful nonfiction accounts of growing up with Holocaust survivor parents was Art Spiegelman's graphic novel *Maus*; serialized between 1980 and 1991, it broke through a cultural barrier, helping others to open up about their suffering. Many psychologists and neuroscientists have examined the traumatized family, finding ever more subtleties, and the story will continue to unfold for decades to come.

An important question is whether epigenetic alterations in stress-related genes, particularly those reflected in the offspring of traumatized parents, are necessarily markers of vulnerability or whether they may reflect a mechanism through which offspring become better equipped to cope with adversity. This is an area we're actively exploring.

It is tempting to interpret epigenetic inheritance as a story of how trauma results in permanent damage. Epigenetic influences might nonetheless represent the body's attempts to prepare offspring for challenges similar to those encountered by their parents. As circumstances change, however, the benefits conferred by such alterations may wane or even result in the emergence of novel vulnerabilities. Thus, the survival advantage of this form of intergenerational transmission depends in large part on the environment encountered by the offspring themselves.

Moreover, some of these stress-related and intergenerational changes may be reversible. Several years ago we discovered that combat veterans with PTSD who benefited from cognitive-behavioral psychotherapy showed treatment-induced changes in *FKBP5* methylation. The finding confirmed that healing is also reflected in epigenetic change. And Dias and Ressler reconditioned their mice to lose their fear of cherry blossoms; the offspring conceived after this "treatment" did not have the cherry blossom epigenetic alteration, nor did they fear the scent. Preliminary as they are, such findings represent an important frontier in psychiatry and may suggest new avenues for treatment.

The hope is that as we learn more about the ways catastrophic experiences have shaped both those who lived through those horrors and their descendants, we will become better equipped to deal with dangers now and in the future, facing them with resolution and resilience.

About the Author

Rachel Yehuda is a professor of psychiatry and neuroscience and director of the Center for Psychedelic Psychotherapy and Trauma Research at the Icahn School of Medicine at Mount Sinai. She is also director of mental health at the James J. Peters Veterans Affairs Medical Center.

Burned Out

By Ulrich Kraft

Since getting his business degree nine years ago, Larry has been the model hard-charging executive bound for the top. After the 28-year-old joined a consulting company, he was quickly promoted to a position with loads of responsibility, a company car and an enviable income. Along with the fulfilling tasks came constant travel and 60- to 80-hour workweeks, including meetings on weekends. But he did not mind. "Occasionally it occurred to me how stressful the job was," he says. "But I really got a kick out of it. For a long time, it was lots of fun."

Until the day he ended up in the intensive care unit. Larry collapsed right outside the door to his apartment, with a terrible headache, a racing heart and vertigo. "At first I thought I had had a stroke," he recalls now, a year later. But the doctor's diagnosis was different: burnout syndrome. The consultant was sick from years of excessive toil.

Larry is not unique, and many experts believe that experiences like his are occurring more frequently in an era of lean staffing. "Perhaps now more than ever before, job stress poses a threat to the health of workers," according to a publication of the National Institute for Occupational Safety and Health, the federal agency responsible for conducting research and making recommendations for the prevention of work-related illnesses and injury. And whereas exhaustion from overwork may happen more commonly in midlife, when energies naturally begin to ebb, it can arise earlier as well. A November 2005 Harris Interactive poll commissioned by Spherion Corporation in Fort Lauderdale, Fla., found that one third of workers ages 25 to 39 already felt burned out by their jobs.

The term "burnout syndrome" was coined in the early 1970s by Herbert J. Freudenberger, a New York psychoanalyst. Freudenberger had noticed that his own job, which was once so rewarding, had come to leave him feeling only fatigued and frustrated. Then he noticed that many of the physicians around him had, over time, turned into depressive cynics. As a result, those doctors increasingly treated their patients coldly and dismissively.

Freudenberger soon began looking at examples outside of health care–and found similar cases in many professions. Afflicted people suffered from mood fluctuations, disturbed sleep and difficulty concentrating. Accompanying the mental distress were physical ailments such as backaches or digestive disorders. Freudenberger defined burnout syndrome as a state of mental and physical exhaustion caused by one's professional life.

No specific statistics track the ailment, partly because burnout syndrome does not have its own classification in the *Diagnostic and Statistical Manual of Mental Disorders*–the bible of the field. Rather it falls under a category of "undifferentiated somatoform disorder." Even without concrete numbers, however, the experts agree: pressure in all trades is rising, and people are struggling to cope.

It is clear that long-term strain plays a central role in burnout. Historically, the body's stress reaction has been a useful protective response. It helps humans–and other animals–survive an immediate threat, such as the appearance of a predator. Before we are consciously aware of it, the brain recognizes a potential danger and sets in motion, in just fractions of a second, a series of physical responses that ready us to fight or flee. Epinephrine, or adrenaline, flows to muscles, in preparation for battle or running. Our senses sharpen. The body shuts down nonessential systems, such as digestion, to conserve energy.

The problem? The alarm swings into action even when the menace is not a hungry-looking bear at the entrance to the cave but rather an irritable boss who wants that PowerPoint presentation finished in 30 minutes. Each time the system gets tripped—as we crunch to meet an impossible production deadline, dash to a meeting, hurriedly pack for yet another last-minute business trip—the adrenal glands secrete stress hormones, the heartbeat speeds up, and blood pressure rises. If such tensions endure for weeks, months or years, physical consequences arise inevitably. Chronic stress contributes to hypertension, heart problems and a weakened immune system, so that we get infections more often.

Vicious Cycle

For many victims of burnout, the fuel for the fire comes from similar sources. It tends to hit the best employees, those with enthusiasm who accept responsibility readily and whose job is an important part of their identity. Larry describes it well: "At a certain point my job had so consumed me that my other needs no longer counted. My overengagement in work led to a constantly worsening state of exhaustion and apathy."

In response to mounting task loads, the wretch piles on the hours, pulling late nights at the office, ignoring exercise, skipping meals or eating unhealthful fast foods on the run, canceling personal dates with friends, missing the kids' soccer games. Ultimately, Larry relates, "I completely isolated myself." Humans are social beings, so we do not fare well when cut off from such networks. "Support from family, friends and colleagues is a vital buffer against stress," says Manfred Schedlowski of the Swiss Federal Institute of Technology in Zurich, which has recently created a research group to study the causes and consequences of work-related stress.

Another risk factor is the level of control a person has over his or her work and the recognition (or lack thereof) that endeavors receive. Juergen Staedt, a psychiatrist who runs the Vivantes Clinic in Berlin, speaks of a woman who was a successful department head for years—until a corporate restructuring during which she was, despite her desperate efforts, unable to prevent layoffs among her employees. It was a slap in the face for her. Plagued by sleep disturbances, loss of appetite and feelings of low self-worth, she ended up at Staedt's clinic. "Such setbacks are a part of life. But people with burnout-syndrome personalities simply can't work their way past them," he explains. "Their entire self-image is shattered." The experts call such stumbles gratification crises–the feeling that their tremendous hard work on the job is not sufficiently noticed–and they add to the problem.

Sooner or later, the ability of these victims to work declines. They find it hard to concentrate, they have few creative ideas, and their memories often fail. They begin to make mistakes. "Then the vicious spiral begins," Staedt explains. "Noticing that you are no longer doing good work increases the pressure on you, and things go from bad to worse." The once generous Larry, for instance, blamed colleagues for his own mistakes and began to criticize and scold them.

Stress and self-dissatisfaction leave their marks on the psyche. Resigned, discouraged, plagued by flagging self-esteem and anxiety about failure, people with burnout syndrome drag themselves painfully through each day. They may seek solace in alcohol or pills. Some even attempt suicide.

The victims are often the last to realize the seriousness of their situation. No one becomes utterly depleted overnight; on the contrary, their batteries run down so gradually that many of them never notice the subtle changes until things are dire. Working long hours, through weekends, they think, "No problem–I am just a little tired right now." But then there is the first time they cancel the tennis game or mumble "sorry" about missing that long-planned weekend trip with friends. The mountain of papers waits on the desk. They cannot leave the work undone, can they?

"Of course, I was aware that things were not going well. But I thought I would deal with it somehow," Larry recalls. When he finally collapsed on his doorstep, he realized at last that he needed help: "That was the shot across the bow that rescued me."

Staedt provides a metaphor for this problem: "If you own a car, you have it inspected each year and you check the oil regularly. Burnout-syndrome patients never bring their cars in for inspection. They drive thousands of miles at full speed and then are shocked when the motor suddenly fails. They have been neglecting routine maintenance."

***IF YOU NEED HELP** If you or someone you know is struggling or having thoughts of suicide, help is available. Call the 988 Suicide & Crisis Lifeline at 988, use the online Lifeline Chat or contact the Crisis Text Line by texting TALK to 741741.*

This inability to find meaning in life outside of work is why advice to just ease up a bit and turn your computer off by five does not work for burnout-syndrome patients. Anyone who hopes to overcome the problem has to learn that satisfaction can come from things other than job success. That is where psychiatrist Staedt comes in. "What we do is something like enjoyment training," he says. "The patients learn how, at last, to do something merely for the pleasure it provides. Baking cookies, painting and taking walks are just as much a part of the therapy as sports and talking to other people."

This is the crux of the matter, Schedlowski believes. "The personality characteristics that have, till now, guaranteed your professional success are what you must now say good-bye to." In treating his exhausted execs, the psychologist can often trace these "master plans" back to the patients' childhoods. "People who, for example, learn as kids always to be punctual, and always do everything perfectly, will profit from that later," he points out. With such ingrained virtues, they did well in school, and later in their careers they rose quickly. Now they also need to master new skills–healthier ones.

This "rewriting the master plan" is the most difficult part of the therapy. It has to do with how the brain functions. Things learned early, and practiced dutifully, become firmly anchored in our brains. "Relearning habits and ways of behavior that are so strongly ingrained is a training process that takes time," Schedlowski states. He recommends six months of ambulatory therapy, during which the patients practice their new, better routines repeatedly in daily life.

Putting on the Brakes

It would be far better never to fall into the vicious cycle of overwork and inner pressure in the first place. The Zurich team is working to counter this eventuality through educational programs. "Stress has become almost normal in today's business world," Schedlowski observes. "If you know how to protect yourself against its effects, the risks of burning out are much lower."

The quantity of stress is a determinant but not the decisive one. "If someone works 12 hours a day, every day, yet still has found a way to relax, he will very likely have no problem," Staedt explains. "On the other hand, someone else may find a part-time job extremely stressful–and develop burnout syndrome."

Rule number one: budget your physical resources. The antistress measures that are involved are as simple as they are effective. They include eating wholesome foods at mealtimes, exercising regularly and getting enough sleep.

Rule number two: workaholics must aim for equilibrium between tension and relaxation. Or, in the language of those in the field, find their work-life balance. "Everyone has to find their own stress-compensation mechanism," Schedlowski says. One woman may reduce stress by running long distances, whereas another may lie on her sofa at home listening to classical CDs, and still a third may tend the rose bushes in her garden. The hobby itself doesn't matter; it is devotion to a pleasurable activity that does.

Close social contacts are also important. Spending time with friends, family, even colleagues at work, protects against excessive stress. Last, it is helpful to learn some relaxation technique, such as yoga or progressive muscle relaxation.

The decisive step, Schedlowski emphasizes, must be made first in your head. "As early as possible in your professional career, you must absorb the idea that physical and mental health are at least as important as climbing the ladder toward success," he says.

Ultimately, Larry made the switch to a more balanced existence. He resigned from his job, repaired his friendships and fulfilled a boyhood dream by taking a round-the-world trip. Then he returned to being a consultant, landing a better spot than the one he had left. But now he balances labor with "more sports, more leisure time, more downtime, more time to enjoy life. Even though my job is, as before, is important to me, these things have priority now." He likes the result. "Things have never been better!"

Referenced

Burn-Out: The High Cost of High Achievement. Herbert J. Freudenberger. Anchor Press, 1980.

Women's Burnout: How to Spot It, How to Reverse It, and How to Prevent It. Herbert J. Freudenberger and Gail North. Doubleday, 1985.

Why Zebras Don't Get Ulcers. Third edition. Robert M. Sapolsky. Owl Books, 2004.

About the Author

Ulrich Kraft, a regular contributor to Gehirn & Geist, *is a freelance science writer in Berlin.*

The Biggest Psychological Experiment in History Is Running Now

By Lydia Denworth

The impact of COVID-19 on the physical health of the world's citizens has been extraordinary. By April 2021 there were almost 130 million cases spread across more than 190 countries. The pandemic's effect on mental health could be even more far-reaching. At one point roughly one third of the planet's population was under orders to stay home. That means 2.6 billion people—more than were alive during World War II—were experiencing the emotional and financial reverberations of this new coronavirus. "[The lockdown] is arguably the largest psychological experiment ever conducted," wrote health psychologist Elke Van Hoof of Free University of Brussels-VUB in Belgium. The results of this unwitting experiment are only beginning to be calculated.

The science of resilience, which investigates how people weather adversity, offers some clues. A resilient individual, wrote Harvard University psychiatrist George Vaillant, resembles a twig with a fresh, green living core. "When twisted out of shape, such a twig bends, but it does not break; instead it springs back and continues growing." The metaphor describes a surprising number of people: As many as two thirds of individuals recover from difficult experiences without prolonged psychological effects, even when they have lived through events such as violent crime or being a prisoner of war. Some even go on to grow and learn from what happened to them. But the other third suffers real psychological distress—some people for a few months, others for years.

Even if most individuals prove resilient, the toll of the COVID-19 disruptions and the sheer numbers involved have experts warning of a mental illness "tsunami." People face a multiple wallop: the threat of disease, loneliness of isolation, loss of loved ones, repercussions of job loss and ongoing uncertainty about when the pandemic will end. Depression, anxiety and post-traumatic stress will undoubtedly follow for some. Mental health hotlines have reported surges in calls, and early surveys found high levels of concern. "This pandemic just ticks all the boxes in terms of the kinds of stressors that are going to be difficult," says psychologist Anita DeLongis of the University of British Columbia, who studies psychosocial responses to disease. The deaths by suicide of health care professionals who had been on the medical front lines are powerful reminders of the risks.

Individual resilience is further complicated by the fact that this pandemic has not affected each person in the same way. For all that is shared–the coronavirus has struck every level of society and left few lives unchanged–there has been tremendous variation in the disruption and devastation experienced. Consider Brooklyn, just one borough in hard-hit New York City. Residents who started 2020 living or working within a few miles of one another have very different stories of illness, loss and navigating the challenges of social distancing. How quickly and how well individuals, businesses and organizations recover will depend on the jobs, insurance and health they had when this started, on whether they have endured hassle or heartbreak, and on whether they can tap financial resources and social support.

The pandemic has laid bare the inequities in the American health care system and economic safety net. Black and Latino Americans are dying at much higher rates than white Americans. "When we talk about preexisting conditions, it isn't just if I'm obese; it's our society's preexisting condition," says medical anthropologist Carol Worthman of Emory University, an expert in global mental health.

Fortunately, the unprecedented pandemic is leading to unprecedented science not just in virology but on mental health and resilience. Behavioral scientists are measuring the psychological toll in real time and striving to identify what helps people cope. Unlike, say, the September 11 terrorist attacks or Hurricane Katrina, which occurred over a finite period even though their effects were drawn out, the open-ended time frame for COVID allows for new kinds of longitudinal studies and research directions. The sudden mass switch to virtual forms of working and socializing is expected to jump-start more nuanced investigations into what makes social interaction satisfying–or stultifying. If researchers meet the challenge of COVID, says psychiatrist Dennis Charney of the Icahn School of Medicine at Mount Sinai, "there will be a whole new science of resilience. We could learn how to help people become more resilient *before* these things happen."

Bend but Don't Break

Rafael Hasid arrived in New York City from his native Israel in 2000 to attend the French Culinary Institute. In 2005 he opened a restaurant called Miriam in Brooklyn that became a neighborhood favorite. In the first weeks of March 2020, Hasid could see what was coming. "I was following the news in Israel," he says. "We were two weeks behind in every respect. I was saying, 'This is going to happen here.'" When Miriam's popular weekend brunch attracted a third of the usual crowd, Hasid did not spend much time wondering what to do: he gave away all of the restaurant's perishable food to the neighbors. By the time the city required all restaurants to shut down, Miriam had already closed.

Faced with potentially traumatic events, "about 65 percent of people are going to show minimal psychological symptoms," says clinical psychologist George Bonanno of Teachers College at Columbia University. Bonanno, who is an expert on resiliency, studies the aftermath of hurricanes, terrorist attacks, life-threatening injuries and epidemics such as the 2003 SARS outbreak. His research and that of others consistently shows three common psychological responses to hardship. Two thirds of people follow a resilience trajectory and maintain relatively stable psychological and physical health. About 25 percent struggle temporarily with psychopathology such as depression or post-traumatic stress disorder and then recover–a pattern known as the recovery trajectory. And 10 percent suffer lasting psychological distress. These results hold true across diverse populations and socioeconomic statuses. "We're talking about everybody," Bonanno says. On the other hand, the risk of psychiatric disorders is twice as high for people on the lowest economic rungs.

But the mental health effects of a crisis so sweeping and insidious may not adhere to this paradigm. Studies show that strict quarantine can lead to negative psychological effects such as PTSD, although few of us have been under true quarantine, which refers to isolating after a possible exposure to infection. Instead much of the world is living with restrictions that Bonanno suspects amount to something more like managing constant stress. "This is the first time in living history we've had a global lockdown that's gone on for such a long time," says epidemiologist Daisy Fancourt of University College London. "We simply don't know how people are going to react to this."

The potential scope of the impact is considerable. "This is different from other forms of stress because it's not just one domain of your life," says health psychologist Nancy Sin of the University of British Columbia. "People are dealing with relationship or family challenges, with financial and work challenges, with health."

Early reports were showing clear effects. The first nationwide large-scale survey in China, where the crisis hit earliest, found that almost 35 percent reported psychological distress. In the U.S., rising fear and anxiety about COVID have been found in people who already suffer from anxiety. Another study captured worrisome findings in older adults. This is surprising because previous research shows that, for the most part, older adults have better emotional well-being. "During this pandemic, older adults don't have those age-related strengths in emotions that we would typically expect," says Sin, who studies aging and is collaborating with DeLongis in an ongoing COVID-19 study of 64,000 individuals worldwide. "They are reporting just as much stress as middle-aged and younger people."

Sin is still analyzing the causes of the stress but suspects it is caused by older adults' higher likelihood of getting sick and of losing loved ones. Older people are coping with their stress better than younger people, however, and reporting less depression or anxiety. They may be benefiting from the perspective that comes with having lived through more than younger people, Sin says. Adults older than 65 have also had more time to develop skills for dealing with stress, and many have retired and so are less likely to be concerned about work.

Fancourt began a study in mid-March 2020 that at one point grew to include more than 90,000 U.K. residents. It is tracking depression, anxiety, stress and loneliness week by week. "We need to know in real time what's happening," Fancourt says. Six weeks in, they found that levels of depression were significantly higher than before the pandemic.

Generally, those with previously diagnosed mental health illnesses, those who live alone and younger people were reporting the highest levels of depression and anxiety. On the positive side, there was a slight decrease in anxiety levels once the lockdown was declared. "Uncertainty tends to make things worse," Fancourt says. Some are frozen by not knowing what is to come, whereas others find ways to carry on.

After Hasid's restaurant had been closed for three weeks, he had not yet received any of the government payments meant to protect small businesses. While his situation was rife with uncertainty, "I was thinking that we have to continue creating business for ourselves," he says. When a few customers e-mailed to inquire if he would consider catering their Passover seders, Hasid developed a prix fixe holiday menu for delivery. Before the pandemic, Hasid was planning to open a delicatessen that would be located in an adjacent storefront. Instead of renovating the new space, he opened the deli inside the restaurant. His biggest worry was whether employees would feel safe. To reassure them, in addition to social distancing, he requires masks and gloves and has someone come in to bleach the restaurant morning and night. Hasid also looked into other sanitizing strategies involving blowers and alcohol that he heard had been used in Singapore.

Hasid recognizes that his ability to adapt is not something every business can do, especially many restaurants that run on tight margins. The new operation used minimal staff, but Hasid continued to pay—out of his own pocket—any employees who were not able to get through to unemployment. Serving food via delivery brings in less than a third of Miriam's former income, but he says it is better than nothing. Last year the restaurant also began preparing a weekly meal for a local hospital. "It is not a money maker, but it's the least we can do." Hasid is pleased with Miriam's reinvention and optimistic that the restaurant will ultimately survive. "We are in a much better situation than a lot of other places in New York," he says.

The Components of Coping

When Brooklyn resident Tom Inck developed a persistent fever and dry cough in the middle of March 2020, the psychotherapist and management consultant feared he had COVID. Because of the shortage of tests at the time, Inck's doctor first screened for every other known virus (Inck paid for the test panel). Then doctor and patient met on the streets of Manhattan. Standing on Madison Avenue in full protective equipment, the doctor administered the test, which came back positive six days later.

Successfully coping in a crisis means continuing to function and engaging in day-to-day activities. One must solve problems (whether that means getting groceries or a virus test), regulate emotions and manage relationships. There are factors that predict resilience such as optimism, the ability to keep perspective, strong social support and flexible thinking. People who believe they can cope do, in fact, tend to cope better.

During nine days of isolation in a spare room, Inck filled the time with meditating and reading. In some ways, things were harder for his wife, Wendy Blattner, who was managing her husband's care, the transition of her marketing agency to remote work, and the emotions of the couple's two college-aged daughters, who were upset at the loss of their semesters and anxious about their father. Blattner left meals outside her husband's door and got up every three hours throughout the night to record his temperature and blood oxygen level. She was scared but resolute. "I felt like he had excellent care, even though it was remote, and that I had the resources within myself and the support I needed," she says. "That's what I told my kids and what I told myself–that it might get rough, but it was going to be okay."

Most people's coping skills can be strengthened. Several of the new studies are designed to identify successful strategies that buffer the effects of the stress. So far, Fancourt says, people are encouraged to follow classic mental health strategies: getting enough sleep, observing a routine, exercising, eating well and maintaining strong social connections. Spending time on projects, even small ones, that provide a sense of purpose helps.

In previous work, DeLongis has shown that those who are high in empathy are more likely to engage in appropriate health behaviors such as social distancing and to have better mental health outcomes than people who are low in empathy. But her earlier studies of diseases such as SARS and West Nile were cross-sectional and captured only a moment in time. Her COVID-19 study followed people's behavior and attitudes for months to capture changes in empathy and coping over time. "This isn't just about a trait of empathy," DeLongis says. Empathetic responses can be learned and encouraged with proper messaging, and her hunch was that increases or decreases in empathetic responding over weeks and months would be associated with shifts in health behaviors and coping mechanisms.

As part of DeLongis's study, Sin asked people to record their daily activities and emotions for a week. In the early months of the pandemic, she says, the picture was "that life is really challenging, but people are finding ways to meet that challenge." Many reported a great deal of positive social interactions, many of them remote. Older adults reported the highest levels of positive experiences in their daily lives, often through providing support to others.

It is striking that remote connections are proving satisfying. Previous research on the effects of digital technology and media focused on the association between time spent on screens and psychological well-being but revealed little about the worth of different kinds of online interaction. Now that the world is relying on the Internet to socialize, investigating those nuances is crucial. Should social media closely mimic face-to-face interaction, or can less intense forms of communication leave people feeling connected? We do not know yet, but it is likely those studies will now get funded when previously they weren't. "I think we just skipped a decade of conversation in a month," says psychologist Amy Orben of the University of Cambridge, who studies adolescent mental health and technology use.

Social media is a factor in other kinds of research as well. Psychologist Roxane Cohen Silver of the University of California, Irvine, is assessing the impact of media exposure on people's well-being. "Those who consume a great deal of news about a community-wide crisis are more distressed," she says. Computational social scientist Johannes Eichstaedt of Stanford University is combining large-scale analyses of Twitter with machine learning to capture levels of depression, loneliness and joy during the pandemic.

As Blattner feared, things did get rough for their family. On nights seven and eight, when Inck's fever hovered around 103 and his blood oxygen levels dropped to 93, his doctor (via Zoom) said if the levels stayed there or got worse, Inck should go to the hospital. "I'm not going to have a patient who dies at home," he said, a statement that alarmed the children. "The toughest thing for us was the fear," Inck says. But Tylenol kept the fever in check, and short, shallow breaths kept Inck's blood oxygen level in the safety zone. After 10 days, he began to feel better.

The experience left Inck grateful and energized. He threw himself back into work counseling others who were sick and signed up to be a plasma donor for critical patients. But, unlike others who recovered, he did not initially venture out much. "The world felt like a vulnerable place," he says.

Society's Preexisting Conditions

Even those brimming with personal resilience need outside help if they face challenges on multiple fronts. As executive director of IMPACCT Brooklyn, a community development corporation that serves the historically Black neighborhoods of Brooklyn, Bernell K. Grier sees just how hard the pandemic has hit the African-American community. "Daily, I'm hearing of people who are either COVID-positive, recovering from it or have died from it," she says. Three of those deaths occurred in apartments that Grier manages and required her to organize deep-cleaning services. Still, she pressed on. "Seniors are fearful of going out, fearful of anyone coming to their front door," Grier says. "They also are not tech-savvy. A lot of things where they're being told to go on the computer, they need someone to hold their hand and help them through the process."

The pandemic, Fancourt says, "is going to exacerbate the social gradient that we're used to seeing across society. It's crucial that [people] have interventions at a national level that can support [them]." In the U.K., such interventions include the National Health Service and a furlough program that pays up to 80 percent of the salaries of millions of Britons who could not work because of the pandemic. In the U.S., paycheck-protection packages and unemployment exist but proved difficult to access quickly.

Grier's organization provides a variety of services around housing, small business advocacy, and interaction with financial and government institutions. As soon as the pandemic hit, her staff distributed information about public health and economic resources. They introduced webinars to help businesses apply for loans. As of late April 2020, "none of the ones that we helped got anything," Grier says. "It's not reaching our businesses." Only 70 percent of Grier's tenants were able to pay rent in April 2020. "We still have to pay the supers, the porters, the heat and electricity, the taxes and everything else," Grier says. "It's a domino effect. If the residents can't pay, we can't pay."

Worthman, the Emory anthropologist, says the ability to cope with the pandemic's reverberations is not just an individual issue but a societal one. It is also an opportunity. "People have pointed to periods of disaster in American history, after World War I and the Depression, that led to real structural change that benefited people."

Grier is advocating for positive change for her community. In her talks with public health and elected officials, she points out disparities such as the fact that the first test centers were not located in poor neighborhoods. "This is a spotlight on what has existed for too long," she says. "When you're looking at [solutions], make sure that income equality and a racial-equity lens is a filter for everything that's put in place." As Brooklyn reemerges from social isolation, Grier knows the critical role groups like hers play. "We will continue to be here to be that liaison, that credit counselor, that navigator."

Cultivating resilience though community support appears to be more important than ever. As a school nurse in Brooklyn, Marilyn Howard, who immigrated from Guyana as a teenager, worked through the early weeks of March 2020 until the public schools closed. She got sick the day after she left work. It took 10 days to get the test results that confirmed she had COVID. By then Howard thought she was on the road to recovery. But on Saturday, April 4, she awoke with labored breathing that rapidly worsened. Her brother Nigel Howard, with whom she shared an apartment, called an ambulance. But April 4 was near the peak of the pandemic in Brooklyn, and there was no ambulance available. Nigel drove them to the nearest hospital, but Marilyn's breathing deteriorated on the way. Less than a minute before they arrived, her heart stopped, and she could not be revived. She was 53.

"A couple of simple things could have saved my sister's life," says Haslyn Howard, the youngest of Marilyn's five brothers. If schools had closed earlier or her colleague could have taken a sick day, she might not have gotten sick. If someone had recommended a pulse oximeter, she would have known to go to the hospital sooner. If an ambulance had been available . . . The Howard brothers arranged a viewing at a Long Island funeral home to provide some closure.

Haslyn permitted only three people in the room at a time, but a simultaneous virtual service allowed more than 250 people to celebrate Marilyn's life.

Nigel later tested positive for COVID and isolated at home. "My brothers and I are in the initial phases of trying to plan an organization that targets efforts to help the Black and brown community, poor communities, address some of these [issues] on a local and tangible level," Haslyn said soon after losing his sister. It is something they can do in her memory that would have made her proud. "That's one of the ways that we're coping," he adds. "How do we turn tragedy into triumph?"

About the Author

Lydia Denworth is an award-winning science journalist and contributing editor for Scientific American. *She is author of* Friendship: The Evolution, Biology, and Extraordinary Power of Life's Fundamental Bond *(W. W. Norton, 2020) and several other books of popular science.*

How Economic Inequality Inflicts Real Biological Harm

By Robert M. Sapolsky

Western cultures have long cherished the notion that all people are created equal. But in the real world, our lives are not balanced with equal opportunities and resources. This distinction was noted mordantly in 1894 by author Anatole France, who wrote that "the law, in its majestic equality, forbids the rich as well as the poor to sleep under bridges, to beg in the streets, and to steal bread." The rich, of course, need none of these things, whereas the poor often have little choice. And economic disparity has only gotten worse during the past several decades, particularly in the U.S. In 1976 the richest 1 percent of U.S. citizens owned 9 percent of the country's wealth; today they own about 30 percent. This trend echoes around the globe.

One of the consequences for the growing poor is worsening health, and the reasons are not as obvious as you might think. Yes, lower socioeconomic status (SES) means less access to health care and living in more disease-prone neighborhoods. And, yes, as the SES ladder's lower rungs have become more populated, the number of people with medical problems has climbed. This is not merely an issue of poor health for the poor and some version of better health for everyone else. Starting with Elon Musk at the top, every step down the ladder is associated with worse health.

The link between socioeconomic inequality and poor health goes beyond simple access to care and living with more dangers. Fewer than half of the health changes along this SES/health ladder can be explained away by risks such as smoking, alcohol consumption and reliance on fast food or by protective factors such as insurance and health club memberships. In the U.K., large, long-term studies of risks in specific groups, led by epidemiologist Michael Marmot and called the Whitehall studies, demonstrated this clearly. Further, this ladder, or gradient, exists in nations with universal health care; if care availability were truly responsible, universal access should make the gradient vanish. Something else, something quite powerful, must be associated with inequities and be able to cause health problems.

That factor seems to be the stressful psychosocial consequences of low SES. Psychologist Nancy Adler of the University of California, San Francisco, and her colleagues have demonstrated that how people rate how they are doing, relative to others, is at least as predictive of health or illness as are any objective measures such as actual income level. The research indicates that poor health is as much about feeling poor as being poor. Epidemiologists Richard Wilkinson and Kate Pickett of the University of Nottingham and the University of York in England, respectively, have filled out this picture in detail, showing that while poverty is bad for your health, poverty amid plenty–inequality–can be worse by just about any measure: infant mortality, overall life expectancy, obesity, murder rates, and more. Health is particularly corroded by your nose constantly being rubbed in what you do not have.

Basically, more unequal societies have worse quality of life. Across countries and among U.S. states, more inequality, independent of absolute levels of income, predicts higher rates of crime, including homicide, and higher incarceration rates. Add in higher rates of kids being bullied at schools, more teen pregnancies and lower literacy. There are more psychiatric problems, alcoholism and drug use, as well as lower levels of happiness and less social mobility. And there is less social support–a steep hierarchy is the antithesis of the equality and symmetry that nourish friendship. This grim collective picture helps to explain the immensely important fact that when inequality increases, everyone's health suffers.

This is where the problem affects the rich, the haves, as well as the have-nots. With increasing inequality, they typically expend more resources insulating themselves from the world underneath the bridges. I have heard economist Robert Evans of the University of British Columbia call this the "secession of the wealthy." They spend more of their own resources on gated communities, private schools, bottled water and expensive organic food. And they give lots of money to politicians who help them maintain their status. It is stressful to construct thick walls to keep everything stressful out.

Knowing that these psychological and social factors influence the biology of disease is one thing. Demonstrating just how these stressors do their dirty work inside the body is something else. How do SES and inequality "get under the skin"? It turns out that researchers have made significant strides toward an answer. We have learned a lot about how poverty affects biology, and the part of the growing inequality gap that worries people is the poverty end. Scientists have been able to trace physiological connections from external inequality to three key inner areas: chronic inflammation, chromosomal aging and brain function.

A Heavy Load

Thinking about the biology of disease was revolutionized in the 1990s, when the late Bruce McEwen of the Rockefeller University introduced the concept of allostatic load. Our bodies are constantly challenged by our environment, and we stay healthy when we meet those challenges and return to a baseline state, or homeostasis. Traditionally this view led scientists to focus on specific organs that solve specific challenges. Allostasis has a different perspective: physiological challenges provoke far-flung adaptations throughout the body. An infected toe, for instance, will produce not only inflammation at the tip of the foot but also wider changes in everything from energy taken from abdominal fat to the brain chemistry of sleepiness. As this biological grind continues, it leads to an array of body parts functioning less than optimally, which can be as damaging to health as a single organ gone very wrong.

Teresa Seeman of the University of California, Los Angeles, took this idea and followed it through the body, measuring various biomarkers of wear and tear, including increases in blood pressure, cholesterol, blood lipids, body mass index, molecular indicators of chronic hyperglycemia, and levels of stress hormones. She showed that this group of disparate measures powerfully predicts physical health and mortality.

Recent research by Seeman and others links low SES with heavy allostatic load because the body is in a constant and futile battle to return to a normal, nonstressed state. These findings highlight an important theme: whereas an adult's SES predicts allostatic wear and tear, childhood SES leaves a stronger lifelong mark. Low SES predisposes youngsters' bodies toward earlier "aging." The scientists also found protective factors. Although growing up in an impoverished neighborhood worsens the low SES–allostatic load link, lucking out with a mother who has the time and energy to be highly nurturing reduces the ill effects.

Stress in any form can produce these effects. It does not have to be related to money, but it is usually related to social situations. My own work with baboons living freely on the East African savanna has shown this effect. In baboon groups, an animal's place in the social hierarchy produces more or less stress. If you are a low-ranking baboon—a socially stressful situation—your body has unhealthy abnormalities in its secretion of glucocorticoids, which are stress hormones such as cortisol. The body also shows unhealthy changes in the gonadal, cardiovascular and immune systems.

In animal and human hierarchies, these stress-induced changes affect health through a key process: chronic inflammation. Few things are better examples of a double-edged biological sword than inflammation. After tissue injury, inflammation contains damage and initiates cell repair. Chronic widespread inflammation, however, causes molecular damage throughout the body, and studies have demonstrated that it contributes to diseases ranging from atherosclerosis to Alzheimer's. Recent work (including my own focusing on inflammation of the nervous system) indicates that chronic high stress levels can promote chronic inflammation. In people, childhood poverty upregulates the adult body's proinflammatory set point, with increased expression of inflammatory genes and increased levels of inflammatory markers such as C-reactive protein, which is associated with a higher risk of heart attacks.

These are long-term effects: more financial losses in the Great Recession predict higher C-reactive protein levels six years later.

Humans share such vulnerabilities with other primates that live in unequal circumstances. Work by Jenny Tung of Duke University shows more markers of chronic inflammation in low-ranking rhesus monkeys compared with the socially dominant animals in a group. Such studies highlight the directness of the link between social stress factors and unhealthy biology because that link occurs in a species that lacks changes in lifestyle risk factors, such as increased rates of smoking and drinking, that we often see in humans who are stuck in low-status situations.

Premature DNA Aging

Progress in understanding the routes into the body taken by the SES/health gradient has also come through a very sensitive measure of aging: the condition of telomeres, which are the stretches of DNA at the very tips of chromosomes.

Telomeres help to keep our chromosomes stable–molecular biologists like to say that they resemble the plastic caps at the ends of shoelaces that prevent fraying. Every time chromosomes are duplicated for cell division, the telomeres shorten; when they get too short, cells can no longer divide, and they lose many of their healthy functions. Telomere shortening is countered by the enzyme telomerase, which rebuilds these tips. Thus, the state of a cell's telomeres tells much about its biological "age," and shortened telomeres that produce frayed, vulnerable chromosomes seem to be a molecular version of wear and tear.

Telomere biology met stress physiology in a 2004 study led by health psychologist Elissa Epel and molecular biologist Elizabeth Blackburn, then both at U.C.S.F.; Blackburn won a Nobel Prize in 2009 for her pioneering work on telomeres. They examined 39 people who lived with severe stress every day: women who were caregivers for chronically ill children. The landmark finding was that white blood cells in these caregivers had shortened telomeres, decreased telomerase activity, and elevated oxidative damage to proteins and enzymes. (Oxidation can disable telomerase.) The longer a child's illness, the more stress the women reported and the shorter their telomeres were, even after the researchers accounted for potentially confounding factors such as diet and smoking. Telomeres normally shorten at a more or less constant rate in people, and calculations showed that these women's telomeres had aged roughly an additional decade–and sometimes more–past those in the low-stress group.

This discovery triggered a flood of supporting studies showing that stressors that include major depression, post-traumatic stress disorder and the experience of racial discrimination can all accelerate telomere shortening. Unsurprisingly, lower childhood SES also predicts shorter telomeres in adulthood; perceived poor neighborhood quality, witnessing or experiencing violence, family instability (such as divorce, death or incarceration of a parent), and other features of poor status early on are tied to these shrunken chromosome tips later in life. Spend your childhood in poverty, and by middle age your telomeres will probably be about a decade older than those of people with more fortunate childhoods.

Thus, from the macro level of entire body systems to the micro level of individual chromosomes, poverty finds a way to produce wear and tear. Most studies of telomere length compare "poor" with "non-poor," as do the studies comparing allostatic load, but the few studies that examine the whole spectrum of inequality, step by low-status step, show that every rung down the SES ladder most likely worsens these biological markers of aging.

Out of Control

Slipping down these rungs also changes the brain and behavior, according to a slew of recent neurobiological studies. My laboratory has devoted more than a quarter of a century to studying what ongoing stress does to the brain in rodents, monkeys and humans. Along with other labs, we have learned that one hotspot is the hippocampus, a region critical to learning and memory. Sustained stress or exposure to excessive glucocorticoids impairs memory by lowering hippocampal excitability, retracting connections between neurons and suppressing the birth of new neurons. In the amygdala, a brain area that is central to fear and anxiety, stress and glucocorticoids heighten those two reactions. Instead of damping things down as they do in the hippocampus, in this fear-promoting region they increase excitability and expand neuronal connections. Together these findings help to explain why post-traumatic stress disorder atrophies the hippocampus and enlarges the amygdala. Another affected area is the mesolimbic dopamine system, which is crucial to reward, anticipation and motivation. Chronic stress disrupts that system, and the result is a predisposition toward the anhedonia of depression and vulnerability to addiction.

Bombardment by glucocorticoids also affects the prefrontal cortex (PFC), key to long-term planning, executive function and impulse control. In the PFC, social stress and elevated glucocorticoids weaken connections between neurons, making it harder for them to communicate. Myelination, the process that insulates cables between neurons and thus helps them pass signals faster, is impaired. Total cell volume in the region declines, and chronic inflammation is activated.

Around the world, health and social problems grow as income disparities widen within societies.

What happens when the PFC is impaired in this way? Lousy, impulsive decisions happen. Consider "temporal discounting": when choosing between an immediate reward and a bigger one if you wait, the appeal of waiting goes down as the time you have to wait goes up. The PFC is normally good at combating this shortsightedness. But stress steepens temporal discounting; the more cumulative stress, the less PFC activation in experiments that call for gratification postponement. For people sliding further into inequality, the less active PFC makes it harder for the brain to choose long-term health over immediate pleasure. That neurological effect might explain why people with more total life stress gain more weight and smoke and drink more than those with fewer stressors.

These changes in the PFC happen in children, too. In separate studies, Martha Farah of the University of Pennsylvania and W. Thomas Boyce, professor emeritus at U.C.S.F., observed that lower-

SES kindergartners typically have elevated glucocorticoid levels, a thinner and less active PFC, and poor PFC-dependent impulse control and executive function. These effects increase as kids get older. By adolescence, lower SES predicts smaller PFC volume. By adulthood, low SES predicts steeper temporally discounted decisions.

Some of these observations present a tricky chicken-and-egg question. The brain changes could lead to poor choices, which in turn lead to deeper poverty, rather than the other way around. But the research suggests that causes and effects run in the other direction, with SES and inequality first influencing PFC function and then other bad things happening.

For example, kindergartners' SES predicted their PFC function; few five-year-olds plummet into poverty by squandering their paychecks on drink and horses. Further evidence comes from a 2013 study by Jiaying Zhao of the University of British Columbia and her colleagues. They examined Indian farmers whose economic fortunes vary seasonably. As individuals' SES went from being poorest during planting season to wealthiest after harvest, improvements in PFC function followed.

To me, the most important evidence comes from research in which people's sense of their SES was lowered by the design of the experiment. Afterward these individuals did heavier temporal discounting. In one 2012 study, subjects played a game of chance against one another, with differing amounts of starting resources. "Poor" subjects became more likely to borrow against future earnings and less attuned to helpful clues about game strategy.

In another study, subjects who were prompted to imagine scenarios of financial loss (versus neutral or advantageous ones) did steeper temporal discounting in an unrelated task. In still other research, subjects were primed to imagine their financial burdens by contemplating an expensive car repair; cognitive function was unchanged in high-SES subjects but declined in poorer individuals.

Why should a transient sense of lower SES induce cognitive changes typical of lower SES in the real world? One explanation is that it is a rational response because it is hard to think about squirreling away money for old age if you can barely buy groceries. Poverty makes the future a less relevant place.

But there is also a powerful stress-related explanation: long-term planning and impulse control tires out the PFC. Increase subjects' cognitive "load" with taxing PFC-dependent tasks, and they become more likely to cheat on their diet. Or you can—and scientists have done this—increase cognitive load by tempting dieting subjects with snacks, and then they do worse on PFC-dependent tests. How much this represents literal "depletion" of the PFC metabolically versus declining motivation is unclear.

Either way, lower SES creates chronic financial worry that distracts and exhausts. It is hard to ace a psychological task of, say, subtracting a series of numbers or a more important task of reining in your drinking when you are worrying about paying your rent. One finding in the car-repair study supports this interpretation. When subjects contemplated a repair of negligible cost, low- and high-SES subjects performed equally well on cognitive tasks.

Of course, we need to better understand the biological consequences of inequality and learn better ways to heal its health scars. But frankly, right now we know quite a bit. We know enough to prompt moral outrage at the situation. It is outrageous that if children are born into a disadvantaged family, they will be predisposed toward poor health by the time they start to learn the alphabet. It should not require us to measure inflammation or the length of chromosomes to prove this is wrong, but if it does, more power to this science.

About the Author

Robert M. Sapolsky is a professor of biological sciences, neurology and neurological sciences at Stanford University and a research associate at the National Museums of Kenya. In his laboratory work, he focuses on how stress can damage the brain and on gene therapy for the nervous system. He also studies populations of wild baboons in East Africa, trying to determine the relation between the social rank of a baboon and its health.

The Stress of Crowds

By Alla Katsnelson

Urban life can be trying—cars and buses honk, passersby jostle, concrete and brick win out over grass and trees. Researchers have known for decades that residents of densely populated areas have higher rates of mental illnesses, including anxiety disorders and schizophrenia. But do the brains of city dwellers function any differently from those of rural folk? Studies are showing that they do.

German researchers recently asked subjects from large cities, small cities and the countryside to undergo a standard psychological stress test—doing arithmetic under time pressure—while having their brain imaged with functional magnetic resonance imaging. Current city living, testing found, correlated with a boost in activity in a brain region called the amygdala, which is associated with memory and emotional intelligence, with a particularly large effect in people from big cities. Even more surprising, subjects who had grown up in a city showed higher activation of a brain area called the anterior cingulate cortex, essentially the amygdala's boss, even if they had later moved to the suburbs or country. The findings were published this past summer in the journal *Nature*. (*Scientific American* is part of Nature Publishing Group.)

Both the magnitude and the specificity of the effect are surprising, says Andreas Meyer-Lindenberg, director of the Central Institute of Mental Health in Mannheim, Germany, and the study's lead author. But he does not yet understand why these brain regions were more active in urbanites under stress. Another recent study suggests that the amygdala and anterior cingulate cortex become activated when one's personal space is invaded. "Maybe it has to do with crowding," Meyer-Lindenberg says.

The activation could reflect the neural machinery involved in managing human interactions, suggests Lisa Feldman Barrett, a psychologist at Northeastern University. She recently correlated amygdala volume with the size of a person's social network. Does a larger or more strongly activated amygdala help you remember new people?

Knowledge of the underlying mechanism should help investigators answer this and other questions more quickly. Traditional epidemiology requires large numbers of subjects to identify broad effects, such as the link between urban life and mental illness. But now researchers can study smaller groups of subjects to see how specific factors—for example, noise in the home or proximity to a green space—play into mental illness and, more broadly, urban stress. Meyer-Lindenberg calls this newer field "neuroepidemiology." That, in turn, could help city planners determine which design features would provide the most solace.

The Myth of Executive Stress

By Keith Payne

It's tough to be the boss. A recent *Wall Street Journal* article described the plight of one CEO who had to drag himself out of bed each morning and muster his game face. It would be a long day of telling other people what to do. It got so bad, we are told, that he had no choice but to take a year off work to sail across the Atlantic Ocean with his family.

Forbes agrees: "many CEOs have personal assistants who run their schedules for them, and they go from one meeting straight to another with barely a moment to go to the bathroom." The indignity! And even worse than the bladder strain is having to fire people: "You may think a CEO can be detached when deciding who to lay off, but generally that couldn't be farther from the truth. Having to make tough decisions about the people all around you can hit very hard." Take heart, those of you who have lost your job in these turbulent economic times. At least you didn't have to fire somebody.

This type of silliness usually cites research from the 1950s on "executive stress syndrome." The research was not on executives, but rhesus monkeys. In a famous experiment, neuroscientist Joseph Brady subjected one group of monkeys to regular electric shocks every 20 seconds for six hour shifts. Another group of "executive monkeys" had the same schedule, except that they could prevent the shocks by pressing a lever in each 20 second period. The "executive monkeys" quickly learned to prevent the shocks by pressing the levers. This situation sounds awful for both monkeys, but decidedly worse for the monkeys with no escape. And yet, it was the "executive monkeys" with greater responsibility and control who started dropping dead from stomach ulcers. These results seemed to suggest that being responsible for making important decisions was so stressful that it posed a serious health risk. Executive stress syndrome was born.

There are of course two problems with an executive monkey: the executive and the monkey. For Rhesus monkeys are not people, and controlling electric shocks is not making business decisions. We can do better.

In fact there are hundreds of studies on the relationship between stress, health, and power. And they virtually all show the opposite of the executive monkeys. Biologist Robert Sapolsky has studied baboon troops in Africa. He finds that the lower the baboon's rank in the pecking order, the more likely it is to have high levels of stress hormones and stress-related illnesses. But a high-ranking alpha male, who can mate with any female he chooses and take out his aggression on any lower ranking male, has much lower stress. If executive apes exist, these are the ones.

Evolutionary psychologists often talk about the brain as a Swiss Army knife, with a particular gadget "designed" by evolution to solve each evolutionary problem. But the stress response is no Swiss Army knife–it's a sledge hammer. The stress response is an all-purpose Code Red that reacts in a similar way to different kinds of threats. The hyena charging from behind the grass elicits the same kinds of bodily responses as the boardroom full of bosses evaluating your PowerPoint presentation. The brain triggers a release of stress hormones including adrenaline and cortisol, causing the heart rate to spike. Glucose floods the system to release energy. All that energy is diverted to muscles in the arms and legs as the body shuts down non-essential activities like growth and digestion. That's great for running or fighting, but no help for remembering your opening joke.

If you are a zebra on the savannah running from the hyena, that adrenaline rush will either save your life or you will become breakfast. Either way, the stress response will be over in a few minutes. But humans are endowed with the ability to remember every flubbed joke in the past and to lie awake in bed imagining, in excruciating detail, the disasters that might await in the months ahead. When the stress response is activated for months at a time, it is toxic. The cardiac workout turns into heart disease, the glucose flood sinks into diabetes, and the overworked immune system gives in to infections. The stress response that evolved to save our lives now threatens it.

How does the stress response work in a human chain of command? A recent study by psychologist Gary Sherman and colleagues provides the most direct test yet of the difference in stress between leaders and followers. They studied full time workers in either business or the military who were taking executive education classes at Harvard's business school. They first classified the participants as either leaders or non-leaders. Leaders were defined as those whose job required them to manage other people. On both surveys of anxiety and biological measures of cortisol, the leaders showed substantially lower levels of stress than the non-leaders. The results were the same in both business and the military. Leadership has its privileges.

These results echo a massive study of British government employees that has been going on since the 1960's. The British civil service has an exquisitely detailed hierarchy with clearly defined job grades all the way from cabinet secretaries down to administrative assistants. People in this study are all employed and they all have health insurance. Nonetheless, physician Michael Marmot has found that each rung down the ladder is associated with more stress-related health problems including the biggest health problem of all, death.

When the executive or the general complains that they are "stressed," we have to pay careful attention to what exactly they mean. They may have more emails in their inbox than they can get to. They may work long hours. But in most cases they can say no to requests and they can decide when and how to deal with challenges. They have much more control over how their lives are arranged than does the secretary who schedules their appointments or the janitor who cleans their office.

People so crave control over their lives that when control is scarce they will manufacture it. In studies by psychologist Aaron Kay and colleagues, people made to feel that they lacked control believed more fervently in a controlling God. They believed also in a controlling government, conspiracy theories, and superstitions.

Someone has to be in control. Lacking control is associated with higher blood pressure, lowered immune function, and a host of stress-related diseases. Control is the essence of power, the linchpin binding status to stress.

So why did the executive monkeys drop dead of ulcers if control protects against stress? It turned out that the study had a fatal flaw. The monkeys were not assigned to be in the executive or helpless groups at random, which is the cornerstone of an experiment. The monkeys who learned how to use the lever to prevent shocks the fastest were "promoted" to executives. Those fast learners may have learned fast because they were especially upset by the shocks. If so, then it was not control that doomed them but their heightened stress response to being shocked. There is a lesson here, and not only in the scientific method. If you are trying furiously to control a situation because you are terrified of what would happen if you don't, you are not really in control at all.

About the Author

Keith Payne is an associate professor in the Department of Psychology at the University of North Carolina at Chapel Hill. He studies unconscious and unintended influences on thought and behavior.

How Much Worry about Mass Shootings Is Too Much?

By Stephanie Pappas

A shopping mall in Texas, a private school in Tennessee, a bank in Kentucky and a dance studio in California: these are the sites of some of the public mass shootings in the U.S. in 2023 alone, representing just a slice of the presumed safe spaces rocked by these tragedies. As mass shootings in the country have risen, evidence is mounting that they are having a far-reaching mental health impact. A 2019 survey by the American Psychological Association (APA) found that 79 percent of Americans reported stress over the possibility of a mass shooting, and 33 percent said fear of a shooting prevented them from going to certain places.

Now new research published on May 23 in *Death Studies* finds that at the far end of this continuum are people whose anxiety about mass shootings has become dysfunctional, or almost a phobia. A few key symptoms—such as hypervigilance, changes in appetite and dreams about mass killings—indicate someone might need mental health support, says the study's author Sherman Lee, a psychologist at Christopher Newport University in Virginia.

"I'm trying to find that line of how much worry is too much worry," says Lee, whose new research validates a mass shooting anxiety scale that could help screen people for this type of dysfunctional dread.

People have a broad range of responses to mass tragedies, says Roxane Cohen Silver, a psychologist at the University of California, Irvine, who studies the effects of collective trauma and was not involved in the new study. "There are certainly people who go about their day-to-day experiences without thinking about mass shootings at all," Silver says. Others, she says, change their behavior dramatically, avoiding certain public places or even homeschooling their children.

The fear could be exacerbated by the frequency of false alarms, says Rachel Rizvi, a business intelligence developer in Denver whose oldest child attends high school. Over the course of her child's freshman year in 2022 and 2023, Rizvi says the school went into full lockdown or closed the campus for security reasons seven times—triggering text messages and e-mail alerts to parents.

"As soon as I get that e-mail, my heart jumps into my throat," Rizvi says, "because there are no details. You have no idea if it's happening at your child's school [or] a school that's close by. You don't know if police are on site. . . . As a parent, you start thinking, 'Did I see my kid for the last time this morning?' Doing that for even an hour feels like forever."

Rizvi says the frequent lockdowns at her child's school are a factor in her family's decision to move to a smaller district, where she hopes smaller schools and more personal attention will reduce the danger. "I know it can happen anywhere," she says, "but I'm just hoping that maybe from a statistical standpoint, the odds are even slightly smaller."

In the APA survey, parents, in particular, were impacted. Twenty-eight percent of those with kids under the age of 18 reported frequent or constant stress about mass shootings, compared with 16 percent of people without minor children. Sixty-two percent of parents said they worried their child might become a victim. Black and Hispanic adults were also more likely than non-Hispanic white adults to express fears that they or someone they know would become a victim or to report stress over mass shootings.

There are no firm numbers on how many people experience so much stress that they lose the ability to function well in daily life—but there are hints that a subset of people struggle a lot. For instance, in a paper published this month in *JMIR Public Health and Surveillance*, researchers reported a spike in calls to a mental health crisis line after the 2022 shooting at an elementary school in Uvalde, Tex. There was also an accompanying increase in conversations about firearms and grief.

Lee's new work suggests that certain symptoms are associated with poor coping. His new screening questionnaire focuses on five symptoms to determine whether anxiety might be affecting a person's daily functioning. One symptom is appetite change when thinking about mass shooting, indicating high levels of stress and fear, Lee says. Another is a physical response such as sweating or a pounding heart when thinking about these events. Such symptoms indicate that a person's "fight-or-flight" response is kicking in, putting the body on high alert.

Dreaming about mass shootings and experiencing hypervigilance are two more signs that a person's mental processes are in "a mode of anxiety," Lee says. Finally, avoidant behavior–such as staying home for fear of a mass shooting–is another sign. Experiencing these symptoms for several days over a two-week period may indicate the need for mental health support, he says.

Fortunately there are highly effective treatments for anxiety, Lee says, including cognitive behavioral therapy delivered virtually. Another strategy for better coping might be to limit media exposure to information about mass shootings. In a 2019 paper published in *Science Advances*, Silver and her colleagues found that people who read or watched a lot of news about the Boston Marathon bombing in 2013 became more distressed, which in turn led them to consume more media during the Pulse nightclub shooting in Orlando, Fla., in 2016, leading to still more worry. The effects of this exposure seem to be cumulative.

"We don't see people becoming habituated," Silver says. "Instead we see that there is increased anxiety."

About the Author

Stephanie Pappas is a freelance science journalist. She is based in Denver, Colo.

News about Racial Violence Harms Black People's Mental Health

By Melba Newsome

For months, Desmond Ellington assiduously refused to watch the videotaped killing of George Floyd. When former Minneapolis police office Derek Chauvin went on trial for his murder, Ellington did not turn on the television until he saw a headline on his smartphone that the verdict would be announced within the hour. It was not that he did not care about what was happening; it was that he cared too much.

For the 37-year-old Black singer and actor, the news accounts and social media videos of racial violence and the killings of unarmed Black people that too often go unpunished were not just demoralizing, they were traumatizing. "It gets to the point where you decide, 'I have to turn off the television because I have my sanity to take care of,'" says Ellington, an Arkansas native who lives in New York. "It's like a sadness, a hopelessness. Those images just keep coming like weeds. You pick one weed, and two more sprout up. So you gonna set the whole yard on fire to kill the weeds? It wears on your psyche."

A growing body of research has documented the detrimental effects of both interpersonal and structural racism. The Centers for Disease Control and Prevention notes that centuries of racism have had a profound and negative impact on the mental and physical health of people of color. Investigators at Columbia University found that experiencing racism can result in traumatic stress linked to negative mental health outcomes, such as depression, anger and low self-esteem. The American Public Health Association calls racism a social determinant of health akin to housing, education, and employment and a barrier to health equity.

A study recently published in the *Proceedings of the National Academy of Sciences USA* adds a new layer to an understanding of the pervasive health effects of racism. Lead author David Curtis of the University of Utah and his colleagues showed that widely publicized anti-Black violence negatively affects the mental health of many Black Americans even if they do not directly experience it.

The study's authors conducted the first nationwide scientific assessment of these media reports using 49 high-profile incidents that occurred between 2013 and 2017. They included media reports of 38 police killings of Black individuals, as well as coverage of about nine legal decisions not to indict or convict officers involved in some of those killings.

Using weekly data from the Behavioral Risk Factor Surveillance System survey, the researchers assessed the mental health impacts on people during this period. Black respondents reported more poor mental health days during the time when two or more of these events happened in the country. Legal decisions not to prosecute or convict the officers involved in the killings were most clearly associated with poor mental health days, Curtis says. This reflects Black people's concern and dismay about what they see as the systemic injustice that follows a much publicized incident and the failure to hold anyone accountable more than their reaction to the initial wrong perpetrated by any single individual.

Those emotions also resonate with personal experiences in which racism is ever present. As a senior operations engineer who is starting a job at Amazon later this month, 26-year-old Adam Williams works in a field in which Black people are dramatically underrepresented. That mirrors his life on a college campus just a few years ago. When he graduated from Purdue University in 2017, just 1.6 percent of the students in the College of Engineering were Black, a statistic that figured into the way he was perceived.

Initially, none of Williams's classmates wanted to be his lab partner, and they seemed shocked when he answered questions correctly in class. "I saw Nazi propaganda scattered across campus detailing how I was genetically inferior," he says of literature disparaging Black people. "I was hastily labeled a delinquent, followed through stores, questioned by police under bogus pretenses and frequently assumed that the only way I could be at Purdue was as an athlete."

For his part, Ellington considers himself lucky to have encountered little overt and direct racism. Yet despite achieving a measure of professional success and traveling the country as a member of the cast of *Hamilton*, he cannot shake the feeling that he is just one wrong move away from becoming another statistic.

"I'm still very aware that it could happen to me," he says. "I have to know how to conduct myself when I'm out. I'm always thinking, 'Am I walking or standing too close to this person?' It's a burden that I have to be in that box so I can get back home to my family. It's so emotionally and physically draining. You don't even realize how much until you see another Black man get shot. I think it's in those moments you realize how much of a burden it is to be Black in America."

Psychologist Vanessa Volpe runs the Black Health Lab at North Carolina State University, where she studies the impact of direct and vicarious racism on physical health. She compares the kind of hypervigilance Ellington talks about–trying to navigate the thicket of structural, systemic and interpersonal racism–to driving a car with the parking brake on.

"Because exposure to racism is chronic, frequent and can happen at any time, you have to be vigilant to protect yourself," Volpe says. "Over time, your body is going to experience this wear and tear that will result in much greater rates of morbidity and earlier mortality for Black folks, compared to white people. A lot of times we think, 'Just get a better car or replace the parking brake,' but that's a Band-Aid. What can we do so that Black people don't even have that exposure in the first place?"

About the Author

Melba Newsome is an award-winning science, health and environmental reporter. She lives in Charlotte, N.C.

Why Do We Panic?

By Scott O. Lilienfeld, Hal Arkowitz

"I was driving home after work," David reported. "Things had been very stressful there lately. I was tense but looking forward to getting home and relaxing. And then, all of a sudden—boom! My heart started racing, and I felt like I couldn't breathe. I was sweating and shaking. My thoughts were racing, and I was afraid that I was going crazy or having a heart attack. I pulled over and called my wife to take me to the emergency room."

David's fears turned out to be unjustified. An emergency room doctor told David, a composite of several therapy patients seen by one of us (Arkowitz), that he was suffering from a panic attack.

The current edition of the *Diagnostic and Statistical Manual* (*DSM*) defines a panic attack as an abrupt and discrete experience of intense fear or acute discomfort, accompanied by symptoms such as heart palpitations, shortness of breath, sweating, trembling, and worries about going crazy, losing control or dying. Most attacks occur without obvious provocation, making them even more terrifying. Some 8 to 10 percent of the population experiences an occasional attack, but only 5 percent develops panic disorder. Contrary to common misconception, these episodes aren't merely rushes of anxiety that most of us experience from time to time. Instead patients who have had a panic attack typically describe it as the most frightening event they have ever undergone.

Research has provided important leads to explain what causes a person's first panic attack—clues that can help ward off an attack in the first place. When stress builds up to a critical level, a very small additional amount of stress can trigger panic. As a result, the person may experience the event as coming out of the blue.

Some people may have a genetic predisposition toward panic, as psychologist Regina A. Shih, then at Johns Hopkins University, and her colleagues described in a review article. The disorder runs in families, and if one identical twin has panic disorder, the chance that the other one also has it is two to three times higher than for fraternal twins, who are genetically less similar. Although these findings do not rule out environmental factors, they do strongly suggest a genetic component.

Panic disorder imposes serious restrictions on patients' quality of life. They may be plagued by a persistent concern about the possibility of more attacks and may avoid situations associated with them. To receive a diagnosis of panic disorder, patients must also worry that they might have another attack where it would be embarrassing (say, in a public setting such as a classroom), difficult to escape (such as when one is stuck in traffic), or difficult to find help (for example, in an area with no medical facilities nearby). Panic disorder accompanied by extensive avoidance of these situations results in a diagnosis of panic disorder with agoraphobia; in extreme cases, sufferers may even become housebound.

From Normal Anxiety to Crippling Fear

What are the roots of such incapacitating attacks? Psychologist David H. Barlow of Boston University, who has conducted pioneering research on understanding and treating panic disorder and related disorders, and others believe that panic attacks result when our normal "fight or flight" response to imminent threats–including increased heart rate and rapid breathing–is triggered by "false alarms," situations in which real danger is absent. (In contrast, the same response in the face of a real danger is a "true alarm.")

When we experience true or false alarms, we tend to associate the biological and psychological reactions they elicit with cues that were present at the time. These associations become "learned alarms" that can evoke further panic attacks.

Both external situations and internal bodily cues of arousal (such as increased breathing rate) can elicit a learned alarm. For example, some people experience panic attacks when they exercise because the physiological arousal leads to bodily sensations similar to those of a panic attack.

Why do some people experience only isolated attacks, whereas others develop full-blown panic disorder? Barlow has synthesized his research and that of others to develop an integrated theory of anxiety disorders, which states that certain predispositions are necessary to develop panic disorder:

- *A generalized biological vulnerability* toward anxiety, leading us to overreact to the events of daily life.
- *A generalized psychological vulnerability* to develop anxiety caused by early childhood learning (such as overprotection from our parents) that the world is a dangerous place and that stress is overwhelming and cannot be controlled.
- *A specific psychological vulnerability* in which we learn in childhood that some situations or objects are dangerous even if they are not.

Panic disorder develops when a person with these vulnerabilities experiences prolonged stress and a panic attack. The first attack activates the psychological vulnerabilities, creating a hypersensitivity to external and internal cues associated with the attack. As a result, even medication containing a mild stimulant can provoke an attack.

Still, there is good news. Two findings in particular can provide reassurance for those with panic disorder. The first is that all panic attacks are triggered by known events, even though the sufferer may be unaware of them. This knowledge can reduce the anxiety associated with the sense of unpredictability. Second, it can be reassuring to learn that a panic attack is a misfiring of the fight-or-flight response in the absence of danger.

Basic research not only has helped us understand panic disorder but also has led to effective treatments. In particular, Barlow and his associates developed panic-control treatment, described in their 2006 book *Mastery of Your Anxiety and Panic*. It involves education about panic disorder and somewhat gradual exposure to the internal and external cues that trigger panic attacks, along with changing the catastrophic interpretations of bodily cues so that they no longer trigger the attacks. This treatment has in most instances surpassed drug therapies for the disorder over the long term.

Referenced

Anxiety and Its Disorders. Second edition. David H. Barlow. Guilford Press, 2002.

A Review of the Evidence from Family, Twin and Adoption Studies for a Genetic Contribution to Adult Psychiatric Disorders. R. A. Shih, P. L. Belmonte and P. P. Zandi in *International Review of Psychiatry*, Vol. 16, No. 4, pages 260–283; 2004.

Mastery of Your Anxiety and Panic: Workbook (Treatments That Work). Fourth edition. David H. Barlow and Michelle G. Craske. Oxford University Press, 2006.

Panic Disorder and Agoraphobia. Michelle G. Craske and David H. Barlow in *Clinical Handbook of Psychological Disorders: A Step-by-Step Treatment Manua*l. Fourth edition. Edited by D. H. Barlow. Guilford Press, 2008.

For a referral to a therapist in your area who uses panic-control treatment or similar treatments, contact the Center for Anxiety and Related Disorders at www.bu.edu/card or the Association for Behavioral and Cognitive Therapies at www.aabt.org

About the Authors

Hal Arkowitz and Scott O. Lilienfeld serve on the board of advisers for Scientific American Mind. Arkowitz is a psychology professor at the University of Arizona, and Lilienfeld is a psychology professor at Emory University.

Section 4: Stress Management and Resilience

Your Response to Stress Improves as You Grow Older

By Daisy Yuhas

No one is a stranger to stress. Decades of research make it clear that major life events, such as the death of a loved one or the start of a new job, can take a lot of our energy and attention. But recently scientists have begun to reveal how smaller daily stressors can create substantial difficulties. David Almeida, a developmental psychologist at Pennsylvania State University, has been following the stressors of daily life in a group of more than 3,000 adults since 1995. Almeida spoke with *Scientific American* contributor Daisy Yuhas to describe a benefit of aging that he has uncovered: stress levels go down and coping skills may actually go up after people pass their 20s. Still, upsetting national or global events can increase our distress.

[An edited transcript of the interview follows.]

Q: You've been tracking people's daily experiences for two decades. How has that shifted your perspective as a psychologist?

A: In my work, I try to characterize a day in the life of an individual. I look at how people use their time, how they experience stressors and positive events, their mood and their physical symptoms. I chart how this changes from day to day—the ebb and flow of daily experiences. So even though I'm a psychologist, my unit of analysis is a day, not a person.

The more I've dug into this work, the more I've begun to see that people actually differ from themselves day to day as much as you differ from somebody else. Our identity isn't just who we are based on the average of our experiences. Our identities may be in the range in our behavior, the extent to which we're going up and down with our experiences.

Q: How do you track daily stressors?

A: We ask people to answer a series of structured questions at the end of every day. Originally we used telephone calls, and now we use web-based approaches. We ask about how they spent their time, their mood, their physical symptoms, who they interacted with, and then we ask a lot of questions about the types of stressors they experienced that day. For some studies, we also collect a sample of saliva, which lets us determine the amount of stress hormones in the body.

With that method, we've worked with a large group of people. I want to acknowledge that the wonderful participants in the National Study of Daily Experiences–which is part of a large-scale investigation called Midlife in the United States–have shared their lives with me for the past 20 years. It's been a privilege to follow them.

Q: You recently published findings from an analysis of 2,845 adults–ages 22 to 77 at the start–over 20 years. In that work, you found that people seem less stressed as they grow older. Can you unpack that?

A: Yes, finally some good news about daily stress! It seems to get a little bit better. We find that younger people report more exposure to stressful events–things they find challenging, upsetting or disruptive–than older people do. So people in their 20s may report stressors on at least 40 to 45 percent of days, but by the time they're in their 70s, that goes down to maybe 20 to 25 percent of days.

In addition, we looked at how much distress people experience or the way they respond to stress. Here we see the same type of pattern, with young adults having higher distress on days with stressors than older people. But around 55 years old, that age advantage–where your response to stress gets better with age–starts to taper off and plateau.

Q: Why is there an age advantage in dealing with stress?

A: I think three reasons could contribute and work together. One has to do with the social roles people inhabit. When you're young, these roles could include being a parent of a young child, starting a job, getting into new relationships. New roles are stressful, as are role conflicts that happen when you have multiple roles going on at once.

A second reason could be that as we grow older, we realize we have only so much life left and want to make the most of it—so we are very motivated to enjoy it.

The third reason, which I am most interested in, is that just by virtue of experiences, opportunities and past stressors, we learn how to deal with them and become more expert in coping with daily stressors as we get older.

Q: Does that explain why research suggests older people are happier than younger ones?

A: As you grow older, you can list all these things you shouldn't be looking forward to: poorer physical health, loss of friends, being sick and cognitive decline. These are not things you would expect to be related to increased happiness. But we see over and over that as people grow older, they have increased life satisfaction.

That said, there is a point when this pattern stops. Much later in life—in someone's 80s or 90s—I think we're seeing a time where things are really tough, and there's a decrease in life satisfaction.

Q: How do things such as economic and political uncertainty affect our day-to-day stress?

A: We were able to study the effects of the 2008 recession and the postrecession period. From our data, it's fairly clear that compared with people in 1995, adults in 2010 had more stressful daily lives and were more upset by their experiences. Our hypothesis is that this reflects historical changes such as the recession and the use of technologies that have changed social interaction. From that, we can speculate on how economic downturn and other changes may affect us. In future work, we hope to see what the pandemic has done—it's possible that we won't see much of an age advantage, for example, in this period.

But what really surprised us from our analysis of the 2008 recession was that this difference in stress seems to be concentrated among people who were in midlife. I'd have thought that younger adults just starting their careers and older adults in retirement would have been worse off, but no: it was adults in their mid-40s through mid-60s who reported higher levels of psychological distress. I think that has to do with the social roles of a midlife adult. They are worried about their kids but also their parents.

Q: On a practical note, should we be trying to remove all stressors from our daily lives?

A: There's something that might actually be good about having some daily stress. People who report having no stress in their lives—you think they are lucky, happy people. But they also report fewer positive things in their lives. They have fewer people in their lives and perform worse on cognitive tests.

It's the reactivity to stress—how you respond to it—that really matters to your health and well-being. It's not the number of stressors but actually your emotional responses that can, for example, raise your risk of cardiovascular disease, increase inflammation and contribute to dying earlier.

Q: How should we manage our responses?

A: There are things people can do, such as eating well and getting enough sleep. But remember that not everyone can do them. It's not all about individual choice.

We've found that minoritized groups—by race, ethnicity and sexual orientation—have higher stress reactivity. They don't always have the resources to cope with daily stressors. For instance, when your body is experiencing stress, it wants to mobilize energy. So getting up and taking a walk is the best way to stem that emotional response. But many people cannot just get up in the middle of their workday and take a walk somewhere.

We need to start talking about how to provide resources to empower people so they can take care of themselves.

About the Author

Daisy Yuhas edits the Scientific American *column Mind Matters. She is a freelance science journalist and editor based in Austin, Tex. Follow Yuhas on Twitter @ DaisyYuhas.*

A Single, Quick 'Mindset' Exercise Protects against Adolescent Stress

By Lydia Denworth

Close your eyes. Cast your mind back to high school and a high-stakes moment in your most difficult course with your toughest teacher. I'll go first: Senior year, Mr. Trice, the final exam in AP Physics. I remember where I was sitting. I remember staring at the paper, feeling I didn't know any of the answers. My heart was pounding; my palms were sweating. I was certain I would fail.

There wasn't a happy ending for me about overcoming adversity. I was able to discard my final test score by taking the actual AP exam, which I also bombed. Not surprisingly, I think of that experience as all bad, an enduring embarrassment. But maybe it didn't have to be that way. Recalling such an experience in a new light is step one of a promising new intervention designed to help adolescents reframe stress and anxiety. Step two is arming them with clear, accessible information so that the next time they feel that way, they will see the experience as a path to learning and growth and even a resource to thrive.

A paper published on July 6 in *Nature* reports that this one-time online intervention, which takes about 30 minutes, improved the way young people thought about stressful events (such as my physics test) and their fretful responses (such as my racing heart). The intervention combines growth mindsets, the belief that ability is not fixed but can be developed with effort and support, and stress-can-be-enhancing mindsets, the belief that physiological responses to stress can be an asset. As students work through the exercise themselves, they read information about how the brain builds on experience and how the body uses stress—a pounding heart delivers extra oxygenated blood to the brain, the better to help you think. They see how it works in the real world: there is a story about a calculus professor who greets students on the first day of class with a reminder of the frustrations they will surely feel and the reassurance that struggle is learning. The intervention hints at strategies for success in stressful situations ("Remind yourself feelings of confusion and struggle when doing difficult schoolwork won't last forever"). And it asks participants to write about what they might do differently next time.

Rigorously tested in multiple experiments involving thousands of high school and college students before and during the pandemic, the brief, scalable intervention appears to shift something fundamental: our interpretation of the world around us and our response to it. Compared with those in a control group, participants in the intervention group thought about stress differently, turning it into a means of energizing the motivation to accomplish a goal. But the intervention also changed their physiological responses for the better, triggering the body to respond to events as a challenge rather than a threat. It lowered cortisol levels and improved cardiovascular functioning. It also lessened overall anxiety levels, with lasting effects in some cases.

"Difficulty and struggle are your friend," says Christopher Bryan, a social psychologist at University of Texas at Austin and an author of the new study. Those experiences don't feel good in the moment, he says, "but it's the path anyone who ever became truly excellent at anything had to travel."

The intervention is not just a hypothetical exercise. The stress that adolescents feel has reached alarming proportions. Last December Surgeon General Vivek Murthy declared adolescent mental health a public health crisis exacerbated by the pandemic, and anxiety disorders lead the mental health challenges faced by young people. "If you can shift your mindset about what anxiety is and what it isn't, how to be anxious in the right way, everything changes," says Tracy Dennis-Tiwary, a professor of psychology at the City University of New York's Hunter College and author of *Future Tense: Why Anxiety is Good for You (Even Though It Feels Bad)*, who wasn't involved in the study. "This paper is a beautiful empirical demonstration of that potential."

Criticisms of some previous mindset research emphasized the lack of statistical rigor or meaningful effects of an intervention on participants. The new paper uses Bayesian analysis, which is widely considered a more reliable measure of the effects of behavioral interventions than other techniques, such as null hypothesis tests of statistical significance. The effect sizes—measuring how strong a finding is—varied from small to large across the six experiments. And as would be expected, they were higher in the laboratory experiments than in the real world. But they were consistently meaningful. "[The study] had broad, multilevel impact on important and well-validated indices of stress and anxiety," Dennis-Tiwary says.

Intriguingly, the intervention did not work for everyone in the same way. "The most vulnerable people in the most stressful time benefit the most," says David Yeager, a developmental psychologist at U.T. Austin and a co-author of the paper. He emphasizes that the intervention is not intended to be used for survivors of trauma and abuse, but administering it broadly does no harm. In addition to addressing mental health issues, a goal of the intervention is to help adolescents engage with challenging courses and projects. In a charter school in one of the experiments, 63 percent of participants passed their math and science classes, compared with 47 percent of students in a control group.

The researchers found that they had to rework a previous growth mindset intervention. That earlier exercise proved effective, especially for low-achieving students, in a national study of more than 12,000 students reported in *Nature* in 2019. But it didn't consider the visceral butterflies-in-your-stomach feeling. "That's a limitation of previous mindset interventions because we forgot about or didn't tap into those stressful emotions," Yeager says.

The usefulness of the new "synergistic" intervention could be considerable, he says, although more study of its lasting effects is warranted. The exercise is currently centered on academic outcomes but could be tailored for use in athletics or in the workplace. It is already used with incoming first-year students in math and science classes at UT Austin. Once it has been more thoroughly tested,

Yeager would like to make the intervention freely available to high schools and colleges nationwide. Another way to scale the idea, he says, is to show professors and managers not only how to use the intervention but also how to support the ideas it explores when they talk to young people entering campus life or the workplace.

These researchers don't just want adolescents to reframe the way they think about stress; they want adults to reframe the way they think about adolescents. "We propose an alternative narrative that emphasizes the role of young people in taking on the formidable challenges of the future," they write in the paper. "Our studies suggest that we might not teach adolescents that they are too fragile to overcome difficult struggles, but that we might, instead, provide them with the resources and guidance that they need to unleash their skills and creativity in addressing big problems."

About the Author

Lydia Denworth is an award-winning science journalist and contributing editor for Scientific American. *She is author of* Friendship: The Evolution, Biology, and Extraordinary Power of Life's Fundamental Bond *(W. W. Norton, 2020) and several other books of popular science.*

The Nature Cure

By Jason G. Goldman

By now it's almost common knowledge that spending time in nature is good for you. Areas with more trees tend to be less polluted, so spending time there allows you to breathe easier. Spending time outdoors has been linked with reduced blood pressure and stress, and seems to motivate people to exercise more.

"So it'll come as no surprise that there's research showing that spending time in nature is good," says University of Exeter Medical School researcher Mathew P. White. "I mean, that's been known for millennia. There's dozens of papers showing that."

"We get this idea, patients are coming to us and they're saying, 'doctor, how long should I spend?' and the doctor is saying, 'I don't really know.'"

So White and his team decided to find out by using data collected from nearly 20,000 people in England through the Monitor of Engagement with the Natural Environment Survey.

And their answer? Two hours a week. People who spent at least that much time amid nature–either all at once or totaled over several shorter visits–were more likely to report good health and psychological well-being than those with no nature exposure.

Remarkably, the researchers found that less than two hours offered no significant benefits. So what's so special about two hours?

"I have absolutely no idea. Really. We didn't have an a priori guess at what this would be, this threshold. It emerged. And I'd be lying if I said we predicted this. I don't know."

Even more noteworthy, the two-hour benchmark applied to men and women, to older and younger folks, to people from different ethnic backgrounds, occupational groups, socioeconomic levels and so on. Even people with long-term illnesses or disabilities benefited from time spent in nature–as long as it was at least 120 minutes per week. The study is in the journal *Scientific Reports*.

While the findings are based on a tremendous number of people, White cautions that it's really just a correlation. Nobody knows why or how nature has this benefit or even if the findings will stand up to more rigorous investigation.

"I want to be really clear about this. This is very early stages. We're not saying everybody has to do 120. This is really to start the conversation, saying, what would a threshold look like? What research do we need to take this to the next step before doctors can have the true confidence to work with their patients? But it's certainly a starting point."

About the Author

Jason G. Goldman is a science journalist based in Los Angeles. He has written about animal behavior, wildlife biology, conservation, and ecology for Scientific American, Los Angeles *magazine, the* Washington Post, *the* Guardian, *the BBC,* Conservation *magazine, and elsewhere. He contributes to* Scientific American's *"60-Second Science" podcast, and is co-editor of* Science Blogging: The Essential Guide *(Yale University Press). He enjoys sharing his wildlife knowledge on television and on the radio, and often speaks to the public about wildlife and science communication.*

How I Broke the Cycle of Stress

By Ashten Duncan

Most of us can likely relate to this experience: you put all that you have into your work for prolonged periods and feel physically and mentally exhausted by the end. This common phenomenon is known simply as stress, a natural reaction to facing increasing demands without the necessary resources to meet them. In moderation, stress on our bodies and minds can elicit beneficial responses that enhance well-being, such as increased memory formation. As most of us know, however, there is often an imperceptibly fine line between advantageous stress and toxic stress from being exposed to too much for too long. Sadly, we often cannot control the balance between the two.

Whether it is on social media or our go-to news channels, we hear a lot about stress and deteriorating well-being in the U.S. In fact, this reality has become so pervasive that some of us probably normalize much of it, writing off the downstream effects of stress as just part of modernity. For adults working in fields associated with above-average levels of psychosocial stress, a vicious cycle can manifest wherein unmitigated chronic stress breaks us down, drains us of our resilience, narrows our perception of what makes life worth living, and leaves us feeling hollow. This entire process is often encapsulated in a single word: burnout.

For some, burnout, schmurnout! Critics of this concept tend to point out how nebulous it sounds even when someone offers up the definition. A different, more integrated way of understanding the long-term effects of chronic stress on the whole person is a principle called allostatic load. Although a lot happens to our bodies when we experience prolonged stress, the central idea of allostatic load theory is that if you apply enough force for long enough on someone without destroying the person, you will fundamentally change the way they function and respond to future stressors, through a process called maladaptation.

As a medical student, I see this problem of maladaptation frequently in both trainees and trained physicians. Research reflects this trend in the alarming number of physicians who quit practicing medicine every year. I think it is because we are so deeply focused on trying to carry out a complex mission of providing care to patients. In our efforts to do this, we often focus on one large goal after another, not allowing ourselves to have much latitude. Whether it is taking a standardized examination or meeting documentation requirements, setting and pursuing less diverse goals in this way is problematic: any barriers to goal attainment will cause much more stress than they would if we had other irons in the fire.

Major news sources and professional organizations have cited emerging evidence of epidemic levels of burnout among physicians in the U.S. This problem is not unique to physicians, but there is perhaps a pronounced susceptibility in this population because of how purpose-discordant tasks such as navigating electronic medical records hinder progress toward meaningful goals such as direct patient care. Although much work is being done to streamline these systems-level processes, not enough is being done to examine what goals are not being achieved and why they exist on a cultural level.

Doing what is in the best interest of the patient is at the heart of the Hippocratic oath, which represents a major theme in physician culture. The oft-overlooked connection between personal well-being and quality of patient care has an empirical basis: physicians who experience chronic stress and burnout are more likely to make errors that result in disease, disability or death. What is ironic is that physicians who push themselves to deliver the best care possible may be doing so at the expense of this goal!

So What Do We Do About It?

We are making strides toward rectifying issues ingrained in our health-care systems, but progress toward shifting the culture to uphold our mission of optimal care is less robust. Research in positive psychology and organizational dynamics has shown that goal diversification is associated with subjective well-being. This means that the act of setting and pursuing different kinds of goals that are meaningful to you is associated with less stress, which decreases the impact of allostatic load on the body and mind. Studies on intervention strategies to promote goal diversification have included mindfulness meditation, hope building and active self-reflection exercises.

Hope theory describes how thinking about our motivation and perceived capability of attaining a goal and about the pathways leading to that goal influences how we set and go after what is important to us. It also suggests that goals, especially smaller ones, can serve as pathways to other, larger goals. This theoretical framework is further supported by research showing that goal setting promotes behavioral change, which can improve well-being even in the context of major depression. Moreover, the act of setting goals, particularly those that are diverse and of high personal value, increases productivity at work, which may subsequently help alleviate stress by decreasing the demands of one's work environment.

One of the most significant hurdles in health care is that clinical practice generally has a substantial workload and numerous legal constraints that force clinicians to stick to a script. So if the current practice environment largely opposes personal goal setting and goal diversity, how do we work around this? National research on health-care quality offers some ideas: decreasing daily workplace activity, adopting collaborative care models and allowing employees to work on quality-improvement projects. A common thread in all of these ideas is that health-care providers can focus on more than just the same repetitive goals, which allows them to feel that they are fulfilling their professional purpose more often. Additionally, this method can limit stress and prevent much of the maladaptation that comes with it.

People can break their cycles of stress and even flourish outside of the work they regularly do by incorporating new and different goals in their day-to-days. Our best shot at getting ahead of the problems stress produces is to look at where our finite energy is going and ask the tough questions: Why? At what cost? We are not machines; it is not sustainable for us to push ourselves at the cost of our well-being. Whether it is in medicine or any other field, we must ensure that our cultures nurture goal diversity and allow us to be more than just our work because, at the end of the day, stress occurs when we feel that we lack the resources to meet the demands around us.

The views expressed are those of the author(s) and are not necessarily those of Scientific American.

About the Author

Ashten Duncan is a fourth-year medical student at the University of Oklahoma–Tulsa School of Community Medicine and a researcher in positive psychology and well-being. Previously he studied interdisciplinary public health.

Enhance Your Resilience

By Steven M. Southwick, Dennis S. Charney

As a college student at Brown University, Jerry White spent his junior year abroad studying in Israel. On a sunny day during the Passover holiday in April 1984, White and two friends set out for a camping trip in the Golan Heights. "I was walking out ahead of my friends with a song in my heart. I like being the leader, the one out in front," he recalls. "Then, boom! A huge explosion." He had stepped on a land mine. As blood poured from his leg, White screamed, "I have no foot! I have no foot!" White's friends wrapped his stump with a shirt, tied a makeshift tourniquet around the injured leg and carried him through what they now knew was a minefield. For the next four months White lived in two Israeli hospitals where he felt helpless, sad and alone. "People were trying to introduce themselves, but they were all missing arms, legs, eyes, or they were burn victims. I felt sick and afraid," he says.

Eventually White returned home, completed college and worked as a substitute teacher before becoming an activist working on behalf of fellow victims. In 1995 he, along with his friend and colleague Ken Rutherford, who had lost both legs to a land mine in Somalia, founded the Landmine Survivors Network (renamed Survivor Corps). This group played a leading role in the International Campaign to Ban Landmines, which was awarded the Nobel Peace Prize in 1997.

White's transformation from frightened victim to passionate survivor demonstrates resilience, the "process of adapting well in the face of adversity, trauma, tragedy, threats or significant sources of stress," according to the American Psychological Association. A resilient person may bend but does not break when confronted with adversity, enabling him or her to bounce back relatively quickly.

Biologically, resilience is the ability to modulate and constructively harness the stress response–a capacity essential to both physical and mental health. Unchecked, chronic stress can impair the immune system and contribute to illnesses such as gastric ulcers, asthma, depression, diabetes and heart disease. Stress can also spawn unhealthy behaviors such as smoking and excessive alcohol use.

Success appears to hinge on resilience. This is especially important during times of prolonged stress and suffering, such as the current COVID-19 pandemic. Setbacks are part of any human endeavor, and those who react to them productively will make the most progress.

Whether a person perseveres or gives up in hard times depends on influences at multiple levels, from molecules to neighborhoods. Resilience is determined by both inborn traits and environmental factors that affect the capacity to adapt to stress. Although some of these environmental influences, such as poverty, are difficult to alter, a person can increase his or her level of resilience by developing mental and physical habits that foster positive adaptation to stress and trauma.

Standing Up to Stress

Browse through the magazine section of any bookstore, and you will find an assortment of articles on how to reduce stress: "7 Rules for a Stress-Free Life," "Leave Your Stress Behind," "How You Can Reduce Your Stress and Your Dress Size," and so on. The basic message is that stress is bad—but it is not always. Some stress is actually good. Without stress, we would weaken. And difficulty that can be mastered facilitates growth, self-esteem, self-efficacy and resilience. A resilient person is thus not someone who avoids stress but someone who learns how to tame it.

A body of data, including studies of identical twins, suggests that certain personal characteristics that foster resilience may be at least partly inherited. These traits include a temperament that leads a child to confidently take on novel tasks and interact with unfamiliar people, as well as a good-natured, sociable personality. Researchers are now uncovering some of the basic biology of resilience that facilitates our adaptive responses to stress.

Beyond biology, a number of environmental variables affect resilience, among them family support, the stability and quality of schools, and the services in and safety of a neighborhood. For example, a loner who is unemployed will have more trouble dealing with stress and trauma than a financially secure individual in a supportive family. Weak leadership in a community and insufficient first-responder and financial resources can also compromise responses to adversity.

Some of these situational factors, along with specific character traits, may be hard to change. Nevertheless, people can learn to think and act in ways that greatly enhance resilience. Virtually anyone can become more resilient through disciplined, consistent practice. The more we activate specific brain areas through our behavior, the more neural connections form in those areas, enabling the neurons involved to transmit their messages more efficiently.

In our book *Resilience: The Science of Mastering Life's Greatest Challenges*, we identify 10 avenues for building resilience. Some of these strategies reduce stress. Others help you grow from the experience. Strategies to increase resilience include learning how to regulate your emotions, adopting a positive but realistic outlook, becoming physically fit, accepting challenges, maintaining a close and supportive social network, and observing and imitating resilient role models. We discuss each of these approaches below.

Emotional Rescue

The ability to regulate emotions such as sadness, anger and fear is critical for effectively coping with stress and trauma. Negative emotions are natural and often adaptive, but if left unchecked, they can seriously compromise our ability to think clearly and to make rational decisions in challenging situations.

Researchers have devised various strategies for regulating emotions and enhancing resilience. Two approaches that have received increasing scientific support in recent years are cognitive reappraisal and mindfulness meditation. Using reappraisal, individuals reinterpret the meaning of an adverse event so that they see it as less negative. Doing so tends to attenuate physiological and emotional reactions to the event. For example, psychologist Kevin Ochsner of Columbia University and his colleagues have shown that when people intentionally reinterpret a situation such as a rejection for a job as being less negative, they report a decrease in unpleasant emotions. Ochsner's team also found that this improvement in mood is accompanied by changes in the brain—in particular, an increase in activity in the prefrontal cortex, a center for planning, directing and inhibiting behavior, and a decrease in activity of the amygdala, a hub for feelings such as fear.

For years researchers have studied how people regulate their emotions. Individuals who frequently use cognitive reappraisal as a way to alter their emotional reactions to stress and trauma tend to report greater psychological well-being than those who do not look for neutral or positive ways to interpret their challenging circumstances. For example, in a 2008 study we, along with psychiatrist Adriana Feder and other colleagues at the Icahn School of Medicine at Mount Sinai, interviewed 30 former Vietnam prisoners of war about how they evaluated their wartime experiences. We found that most of these veterans, many of whom had been brutally tortured, had actively reappraised their imprisonment and found meaningful ways in which they had grown stronger, wiser and more resilient as a result of it. They also reported that they were now better able to see possibilities for the future, relate to others and appreciate life.

Training in cognitive reappraisal is a component of various therapies designed to enhance well-being, strengthen resilience and reduce distress. For example, cognitive-behavioral therapies, which are effective for treating mood and anxiety disorders, teach individuals to observe their own thoughts and behaviors, to challenge their negative assessments of stressful situations and of themselves, and to replace these with more realistic and positive points of view. When faced with a highly stressful or negative episode, you may find it helpful to ask yourself a few questions: "Is there a less destructive way to look at this situation?" "Am I catastrophizing or exaggerating its potential negative impact?" "Is there something that I can learn from this experience, or is it possible to grow stronger as a result?

Another strategy for controlling stressful emotions is to engage in mindfulness meditation. This exercise teaches practitioners how to consciously live in the present moment rather than dwelling in the past or fretting about the future. The participant becomes an observer who learns to watch, but not judge, as the mind tends to automatically follow familiar conditioned patterns of thinking that often add to distress and maladaptive coping. Mindfulness meditation has been associated with improved ability to focus; more flexible thinking; greater psychological well-being; and better ability to cope with depression, anxiety and stress.

Brain-scanning experiments have revealed that both mindfulness meditation and training in cognitive reappraisal can increase activation of the left prefrontal cortex (a pattern associated with greater emotional control), which provides a boost in positive emotions and faster recovery from feelings such as anger, disgust and fear. Attending a workshop on mindfulness meditation is one good way to start learning how to modulate the stress response and successfully weather life's many challenges.

A Glass Half Full

In addition to attenuating negative emotions, a dedicated effort to bolster positive emotions can enhance a person's ability to bounce back from difficulty. Optimism and positive emotions are strongly associated with good mental and physical health, even longevity. In a remarkable study published in 2001 developmental psychologist Deborah D. Danner of the University of Kentucky and her colleagues examined autobiographical sketches composed by 180 nuns from the School Sisters of Notre Dame in Milwaukee before they took their vows of commitment to the Church and God. They found that the degree of positive emotion expressed in the sketches, which had been written decades earlier, predicted longevity. Only 34 percent of nuns whose sketches were classified in the lowest quarter of cheerfulness were still alive at age 85, compared with 90 percent who had been classified in the most upbeat quarter.

The health benefits of positive emotions probably stem from a better ability to regulate the stress response. Psychologist Barbara L. Fredrickson of the University of North Carolina at Chapel Hill and her colleagues have found that negative emotions tend to increase physiological arousal, narrow our focus of attention and restrict our behaviors to those essential for survival. Positive emotions, in contrast, have been found to reduce arousal and broaden our focus, leading to more creative and flexible responses to stress and trauma.

Resilience is associated with realistic optimism, not the rose-colored form. Because the latter often involves ignoring negative information, people who adopt an overly buoyant outlook tend to underestimate stressful and risky situations, whereas realistic optimists filter out unnecessary negative information but pay close attention to bad news that is relevant to dealing with adversity. To enhance optimism, focus on what is positive around you, intentionally think positive thoughts, embrace humor, reframe the negative in a more positive light, behave in ways that are known to be associated with positive emotions, and try to associate with positive friends and colleagues—emotions can be contagious.

Working Out Your Troubles

Physical activity can enhance resilience by protecting people (and animals) against the negative effects of stress. During the past two decades neuroscientist Benjamin N. Greenwood of the University of Colorado Denver and his colleagues have published a series of studies in rats showing that weeks of voluntary wheel-running exercise can ward off anxiety and depressionlike behaviors, such as exaggerated fear, reduced social exploration and increased intake of morphine, that can result from exposure to a variety of intense stressors.

In humans, aerobic exercise also has been shown to reduce symptoms of depression and anxiety. It improves attention, planning, decision-making and memory, all of which are important for effectively managing stress. Exercise appears to promote resilience through a number of neurobiological mechanisms. For example, it boosts levels of endorphins and of neurotransmitters such as dopamine and serotonin that may reduce symptoms of depression and elevate mood. It also suppresses the release of the stress hormone cortisol.

In addition, workouts seem to activate genes encoding proteins such as brain-derived neurotrophic factor (BDNF) that promote the growth and repair of neurons, which prolonged stress can damage.

In a study published in 2011 psychologist Arthur F. Kramer, then at the University of Illinois at Urbana-Champaign, and his colleagues found that moderate-intensity aerobic exercise (three days a week for one year) increased the size of the hippocampus, a brain region that is involved in both memory and stress regulation, by 2 percent. This growth was associated with increases in BDNF levels and improved memory, suggesting that exercise can protect neurons in this area and may therefore help people recover from difficult circumstances.

For general health, the Department of Health and Human Services has recommended at least one hour and 15 minutes a week of intense aerobic workouts such as swimming or running or twice as much moderately intense exercise–mowing the lawn, say, or walking fast–along with two days of muscle strengthening. (Those who work out more tend to reap greater health benefits, although very intense or prolonged athletic endeavors can actually be detrimental to physical and mental health.)

To enhance resilience, however, we recommend that you consult with your physician and then develop a schedule in which you gradually increase the intensity of your cardiovascular and strength training. Every workout should be challenging but manageable. In this way, in addition to getting the biological benefits of exercise itself, you are also carefully calibrating your exposure to stress in a way that is known to increase resilience.

This general strategy is called stress inoculation. It is based on the notion that if a person deliberately takes on increasingly difficult challenges, he or she will gradually learn to handle higher levels of stress. The inoculation principle of graded exposure can apply to a broad range of activities designed to increase physical, emotional and cognitive resilience. For example, practitioners of mindfulness can increase their ability to concentrate by gradually lengthening their periods of meditation. A person who is afraid to give speeches might sign up for a public speaking workshop and afterward accept speaking engagements in front of small, nonthreatening audiences, then gradually agree to stand up in front of larger and more challenging crowds.

Such experiences should be outside your comfort zone but not so intense as to be unmanageable or potentially harmful. You should also plan to progressively increase the intensity or difficulty of these endeavors. As noted in the *U.S. Army Combat Stress Control Handbook*, "to achieve greater tolerance or acclimatization to a physical stressor, a progressively greater exposure is required. The exposure should be sufficient to produce more than the routine stress reflexes.... In other words you must stress the system." Moreover, make training as realistic as possible. Pilots, firefighters, police and soldiers all enhance their physical, cognitive and emotional skills in realistic training settings where they can obtain constructive feedback from instructors and colleagues.

As you acclimate to greater levels of stress, you will need to recover and relax more. Psychologist James Loehr, who has written books on resilience training, points out that insufficient recovery relative to the high volume of stress can be quite damaging. One of the most important life skills may be knowing when high recovery needs to balance high stress.

Friends in Need

A particularly effective way to enhance resilience is to strengthen your relationships because doing so can dampen your biological response to stress and bolster your courage in tense situations. High levels of social support have been associated with better psychological outcomes after many types of trauma, including childhood sexual abuse and combat. High social support has also been linked with protection against depression and stress disorders such as PTSD, as well as greater emotional well-being and longer life. In contrast, social isolation has a negative impact on psychological and physical health, with effects on longevity similar to those of a sedentary life style, obesity or cigarette smoking.

Backing from others is powerful because it increases an individual's self-confidence, provides a safety net if we should fall and bolsters our belief that we can overcome obstacles. As a result, we tend to more actively solve problems rather than passively avoiding challenges. After all, bonding with others does provide greater security; a group that works closely together is stronger than an individual. On a biological level, social ties activate the parasympathetic nervous system, which lowers subjective and physiological measures of stress and prompts the release of the hormone oxytocin. Oxytocin is known to reduce anxiety and fear, in part by limiting the cortisol response to stress. (It also promotes affiliative behaviors and emotions, such as trust, that encourage continued socializing.)

Interestingly, social support may have an even stronger effect on coping with adversity in Latin American and other cultures that place great value on interdependence, extended kinship systems and collectivism. It plays a comparatively lesser role in more individualistic cultures, although relationships in all societies are vital.

To boost resilience through your relationships, first evaluate your social network. Make a list of the people you feel connected to. With whom do you interact daily? Who would help you without hesitation? To whom do you turn for advice? Who really understands you? Resilience is strongly associated with altruism and giving support to others.

If you find you need to bolster or simply maintain a supportive network, reach out to family, friends and colleagues. Join organizations that share your interests and concerns. Find ways to help others. As an adult, it is also important to teach kids the skills needed to become socially competent, such as knowing how to listen to and empathize with others.

More than Flattery

In your social circles, look for people who recover quickly from hardship whom you could learn from. Members of your own family, colleagues, teachers, coaches, or even historical figures or fictional characters could also serve as resilient role models.

Psychologist Albert Bandura, a highly influential emeritus professor at Stanford University, believes that modeling is most effective when the observer has analyzed what he or she wishes to imitate by dissecting it into segments and creating rules that could guide future action. For example, when a friend lost her job, we noticed that she immediately called fellow employees, former employers and friends to solicit their help in searching for a new position. When she fell ill with pneumonia, she moved in with her sister until she recovered. And when her mother died, she called her family and closest friends and asked them to spend time with her. Rather than going it alone in times of stress, she reached out for support from those who care about her, a pattern that can guide our own behavior. Other resilience-enhancing attitudes and behaviors include perseverance and endurance, moral character and virtues, toughness tempered by compassion, altruism, dignity in the face of deprivation and potential growth from suffering.

The resilience-enhancing strategies described above can be powerful tools for dealing with life's challenges, including the COVID-19 pandemic. This pandemic has caused unprecedented levels of stress, trauma and depression among the general public and among frontline health-care workers, who have risked their own lives to save the lives of their patients.

White, the land-mine survivor, is an outstanding model of resilience. Through years of soul-searching (read: cognitive reappraisal), he began to see the loss of his foot as more opportunity than horrendous misfortune. Together with a group of like-minded individuals, he embarked on a quest to rid the world of land mines. Through thoughtful reassessment, social support and a problem-solving approach, White transformed his tragedy into a mission of hope.

About the Authors

Steven M. Southwick is Glenn H. Greenberg Professor Emeritus of Psychiatry, PTSD and Resilience at the Yale University School of Medicine and the Yale Child Study Center.

Dennis S. Charney is Anne and Joel Ehrenkranz Dean and a professor of psychiatry, neuroscience, and pharmacology and systems therapeutics at the Icahn School of Medicine at Mount Sinai. He and Southwick are authors of Resilience: The Science of Mastering Life's Greatest Challenges *(Cambridge University Press).*

Vision and Breathing May Be the Secrets to Surviving 2020

By Jessica Wapner

We are living through an inarguably challenging time. Earlier this year the U.S. faced its highest daily COVID-19 case counts yet. Uncertainty and division continue to dog the aftermath of the presidential election. We are a nation and a world under stress.

But Andrew Huberman, a neuroscientist at Stanford University who studies the visual system, sees matters a bit differently. Stress, he says, is not just about the content of what we are reading or the images we are seeing. It is about how our eyes and breathing change in response to the world, as well as the cascades of events that follow. Both these bodily processes also offer us easy and accessible releases from stress.

Huberman's assertions are based on both established and emerging science. He has spent the past 20 years unraveling the inner workings of the visual system. In 2018, for example, his laboratory reported its discovery of brain pathways connected with fear and paralysis that respond specifically to visual threats. And a small but growing body of research makes the case that altering our breathing can alter our brain. In 2017 Mark Krasnow of Stanford, Jack Feldman of the University of California, Los Angeles, and their colleagues identified a tight link between neurons responsible for controlling breathing and the region of the brain responsible for arousal and panic.

This growing understanding of how vision and breathing directly affect the brain—rather than the more nebulous categories of the mind and feelings—can come in handy as we continue to face mounting challenges around the globe, across the U.S. and in our own lives. *Scientific American* spoke with Huberman about how it all works.

[An edited transcript of the interview follows.]

Q: What is stress?

A: Stress is one position along the continuum of what we call autonomic arousal. At one end of this continuum would be somebody in a coma. At the very other end of that continuum is a full-blown panic attack: heart racing, pupils dilating, hyperventilating. In between we have lower levels of stress [and the states of being] alert and focused, sleepy and asleep. Stress is generally a high level of autonomic arousal. It was designed to be a generic response to mobilize the body.

Sometimes that's well matched to the demands of life. If you need to run and catch your train, you want all the things that go along with stress to go pursue that train. But if the stress response is spontaneous or excessive, it can start to feel pathological.

Q: What is stress's relationship to vision?

A: When you see something exciting or stressful—a news headline, a fraudulent credit-card charge—heart rate increases; breathing increases. One of the most powerful changes is with vision. The pupils dilate, and there's a change in the position of the lens in the eye. Your visual system goes into the equivalent of portrait mode on a smartphone. Your field of vision narrows. You see one thing in sharper relief, and everything else becomes blurry. Your eyeballs rotate just slightly toward your nose, which sets your depth of field and focus on a single location. This is a primitive and ancient mechanism by which stress controls the visual field.

Q: How does this visual mode affect the body?

A: This focal vision activates the sympathetic nervous system. All the neurons from your neck to the top of your pelvis get activated at once and deploy a bunch of transmitters and chemicals that make you feel agitated and want to move.

Q: Why is the visual field so connected to this brain state?

A: Something that most people don't appreciate is that the eyes are actually two pieces of brain. They are not connected to the brain; they are brain. During development, the eyes are part of the embryonic forebrain. Your eyes get extruded from the skull during the first trimester, and then they reconnect to the rest of the brain. So they're part of the central nervous system.

Having the eyes outside the skull orients the organism to the time of day. But it also means that you've got two pieces of brain that can register events in the environment at a distance in order to adjust the overall state of alertness in the rest of the brain and body. It would be terrible if we had to wait until things were in contact with us before we could prepare to react to them.

Q: Is there a visual mode associated with calmness that can change our stress levels?

A: Yes: panoramic vision, or optic flow. When [you] look at a horizon or at a broad vista, you don't look at one thing for very long. If you keep your head still, you can dilate your gaze so you can see far into the periphery–above, below and to the sides of you. That mode of vision releases a mechanism in the brain stem involved in vigilance and arousal.

We can actually turn off the stress response by changing the way that we are viewing our environment, regardless of what's in that environment.

Q: You are also researching breathing as a way to regulate autonomic arousal.

A: Yes. Vision and breathing are, without question, the fastest and most obvious ways to control autonomic arousal. The way we breathe impacts our states of stress very strongly.

Data show that during sleep and claustrophobic states, people and animals generate what are called physiological sighs–double inhalations followed by exhalations. Children also do this when they are sobbing. A physiological sigh, two or three times, is the fastest way that we are aware of to bring the level of autonomic arousal back down to baseline.

Q: Why does this breathing pattern work to reduce stress?
A: Our lungs consist of tons of tiny little sacs of air—millions of sacs of air. As we get stressed, these little sacs collapse. They deflate like a balloon. Physiological sighs cause the sacs to reinflate. Carbon dioxide is the trigger for breathing: We don't breathe because we need oxygen. We breathe because carbon dioxide levels get too high. Physiological sighs offload the maximum amount of carbon dioxide.

Q: How are you studying the link between breathing and stress?
A: David Spiegel, associate chair of psychiatry at Stanford, and I are leading a study of breathing in which 125 participants have been wearing wrist monitors that measure breathing, sleep duration, heart rate variability and heart rate. The participants are divided into four groups of different breathing modalities: meditation for five minutes a day; repeated physiological sighs; box breathing (equal durations of "inhale, hold, exhale, hold," repeated for five minutes); and deliberate hyperventilation repeated a few times. We want to see which patterns of breathing most rapidly reduce the stress response. We're analyzing the data now.

Q: How are breathing and the brain connected?
A: The relationship is anchored through the diaphragm, the only organ in the body that is skeletal muscle designed for voluntary movement. You can immediately take control of the diaphragm. So breathing represents a bridge between the conscious and unconscious control of the body.

When you inhale, the diaphragm moves down, and the heart gets a little bigger because it has more space. Blood flows a little more slowly through the heart under that condition. So the heart then signals to the brain, and the brain says, "Oh, we'd better speed up the heart." So if you want to increase your heart rate, you inhale more than you exhale. The opposite is also true. Every time you exhale, you're slowing down the heart rate.

Q: So with vision and breathing, you are looking at physiological processes that are automatic but that we can also control.

A: Yes. If I make you stressed, you'll perspire. But you wouldn't say, "I'm going to make myself sweat, and therefore I'll be stressed." You can't control your heart rate directly. You can't control your adrenals with your mind. But you can control your diaphragm, which means you control your breathing, which means you control your heart rate, which means you control your alertness. You can control your vision, which thereby controls your level of alertness, your level of stress and your level of calmness.

Vision and breathing are essential as levers or entry points to autonomic arousal because they are available for conscious control at any point.

About the Author

Jessica Wapner is a science journalist. She is author of Wall Disease *(2020) and* The Philadelphia Chromosome *(2014), both published by The Experiment. Wapner lives in upstate New York.*

Stress Relief Can Be the Key to Success in School

By Clancy Blair

Stress can be toxic at any age. It rattles us when it strikes, shaking up our relationships and narrowing our focus. When it becomes chronic, it ravages our health. Physically, emotionally and intellectually, stress can drag us down.

An even more insidious effect is the assault it can launch on a child's brain, impeding the development of critical cognitive skills. A number of researchers, including myself, have discovered that psychological stress affects the thinking skills and brain development of even very young children, likely beginning prenatally. It is no mystery that stress thrives in difficult situations, but research is now showing that a disadvantaged upbringing may set back children in profound, lasting ways. In fact, stress may be one important mechanism through which poverty adversely affects children's ability to perform well in school.

Although children differ in their susceptibility to the problems of poverty, data show that youngsters from lower-income homes are very likely to start school behind their more affluent peers. This socioeconomic gap persists throughout the school years and is difficult to close. People have long argued that disadvantaged homes tend to offer an impoverished learning environment that does not sufficiently prepare children for the rigors of school. This theory is, at best, only half of the story. My work suggests that the stresses that accompany low income—such as crowded conditions, noise, financial worries and an inability to provide adequate child care—directly impair specific learning abilities in children.

A stressful childhood may emerge from conditions other than poverty, whether from challenging family circumstances such as a divorce or death, overbearing or distracted parents, or learning disabilities that create anxiety about school. A focus on reducing stress through changes in home or in school could improve the well-being of large numbers of schoolchildren and set them up for greater success throughout life.

Flooding the Engines

Stress hormones can shape the developing circuitry of the brain. Most notably, they influence the neural connections in the prefrontal cortex, located behind the forehead, that buttress what are known as executive functions. These include the ability to hold information in mind (working memory) and to inhibit automatic or impulsive responses to stimulation. Executive functions are critical for reasoning, planning and problem solving and for regulating emotions and attention. They are essential to academic success.

The effects of stress on the brain depend on how much of it is present. A little stress heightens alertness; it *improves* people's performance on complex tasks. But as the dose exceeds a certain level, stress starts to erode performance. This relation between arousal and performance can be expressed as an inverted U-shaped curve, first identified by psychologists Robert Yerkes and John Dodson in 1908. In the brain, moderate amounts of stress hormones such as cortisol and noradrenaline boost activity in prefrontal areas that underlie executive functions. At high levels, however, they flood this engine of self-regulation, shutting it down. Over time the brain circuits that control stress hormone levels are shaped by experience toward a tendency to unleash either very large or very small amounts of these hormones onto the prefrontal cortex in response to stress or to maintain a more optimal level of arousal.

In 2001, I began to wonder how this physiology played out in the brains of young children and whether it might explain how poverty "gets under the skin." I set out to explore whether the chronic stress of poverty might be impairing the developing executive function of children enough to set them back at school.

My graduate student at the time, Rachel Peters, and I, then at Pennsylvania State University, gave two tests to 170 four-year-olds in central Pennsylvania enrolled in Head Start, the federal preschool program for children in poverty. We measured mental flexibility by asking children to identify different ways in which small groups of objects were similar. We examined working memory and inhibitory control by asking them to tap a peg twice when the experimenter tapped it once, and vice versa. Here the child has to remember the rule and control the impulse to copy the experimenter. We also asked teachers to rate each child's behavior and academic abilities. And in collaboration with neuroendocrinologist Douglas Granger, also then at Penn State, we took samples of the children's saliva to determine levels of cortisol at the beginning, middle and end of our experimental session.

The children with better executive function and behavior had low cortisol at the beginning of the session that rose and then returned to baseline, as expected, in response to the mild stress of meeting one of us and participating in our tasks. Those who showed either a sustained high level of cortisol or a blunted response—high initial levels, which then dropped, indicating a shutting down of the process—tended to have low executive function; their teachers also rated them as more aggressive and lacking in self-control. We published these results in 2005.

As we followed these children into kindergarten, we observed that executive function matters for achievement: this suite of mental skills was the main determinant of math proficiency, far outweighing other aspects of intelligence. And in an analysis published this year our team, led by postdoctoral researcher Daniel Berry, found that elevated cortisol in children directly predicts academic difficulties, as indicated by their knowledge of math, letters and words. A statistical analysis also backs up our hypothesis that detriments to executive function—as opposed to, say, low general mental ability—are the critical link tying high cortisol to low academic ability.

Positive Parenting

Meanwhile my colleagues and I also set out to determine what aspects of poverty might contribute most to children's stress. We focused on parenting style. Impoverished parents tend to elicit obedience through discipline rather than encouraging exploration and learning by doing. The latter approach, known as scaffolding, is essential to sensitive parenting. In this type of parenting, mothers and fathers interact with their children during play and create opportunities for them to accomplish small tasks, such as stacking blocks. Although poor parents can and do provide sensitive care, they are less likely to do so, given the realities of their situation and, potentially, their own high stress levels.

To investigate further, we have been following 1,292 children, starting at birth, and their families, most of whom live in poverty in rural communities in Appalachia and the Deep South. For about seven years now our team has been visiting these homes annually to collect data on family and economic conditions as well as on executive function and cortisol levels. My colleagues Martha Cox and Roger Mills-Koonce of the University of North Carolina at Chapel Hill analyzed video recordings of mothers interacting with their children in free play. In our first analysis, published in 2008, we found that infants whose mothers displayed the sensitive, scaffolding parenting style had lower cortisol levels and were calmer and more attentive than those whose mothers either completed the activity for them or restricted their attempts to do so.

Furthermore, at age seven months the children whose parents displayed the positive parenting style were more likely to exhibit a healthy cortisol response–a rise and fall–to fear (triggered by an experimenter in a mask) and frustration from a toy placed just out of reach. At age 15 months these children again had lower cortisol levels and were more likely to respond appropriately to the emotional challenges. We now had evidence that parenting style shapes the developing stress response system.

We next sought to sketch the complete path from poverty to parenting to increased stress and diminished executive function in the same group of children. Most recently, we found that the more severely impoverished the family, the less likely parents were to be sensitive and responsive. As expected, the children in such homes had elevated cortisol, which was, in turn, associated with lower executive function. We also saw that less positive parenting went hand in hand with poorer executive function in children, indicating that mothers and fathers can directly stimulate the development of important mental skills.

Creating Capable Kids

Research indicates that stress from a variety of sources—chaotic and poorly run classrooms, for example, or problems with family or peers—impedes learning. The potential good news: knowing that stress is a malevolent force means that finding ways to thwart it could boost children's learning capacity.

In that vein, my collaborators and I are testing a program that teaches parents how to be more sensitive and how to structure opportunities for their children to learn while providing warm and loving care. We are also trying out a new curriculum that gives kindergarteners and preschoolers more control over their learning activities. After a year we will see whether the children's stress regulation and executive function improve. Although this work is in its early stages, we are encouraged by the possibility that informed changes to environments can boost children's self-control and academic competence, giving many of our youth a far greater chance of succeeding in life.

Referenced

Stress Signalling Pathways That Impair Prefrontal Cortex Structure and Function. A.F.T. Arnsten in *Nature Reviews Neuroscience*, Vol. 10, pages 410–422; June 2009.

A Bidirectional Model of Executive Functions and Self-Regulation. Clancy Blair and Alexandra Ursache in *Handbook of Self-Regulation: Research, Theory, and Applica-*

tions. Second edition. Edited by Kathleen D. Vohs and Roy F. Baumeister. Guilford Press, 2011.

Child Development in the Context of Adversity: Experiential Canalization of Brain and Behavior. C. Blair and C. C. Raver in *American Psychologist*, Vol. 67, No. 4, pages 309–318; May-June 2012.

About the Author

Clancy Blair is professor of applied psychology at New York University, where he studies the development of self-regulation in early childhood. He earned his Ph.D. in 1996 from the University of Alabama at Birmingham. His research is supported by the National Institutes of Health.

GLOSSARY

acute Experienced severely or intensely.

agoraphobia The fear of being in situations or places that may cause feelings of helplessness, embarrassment, or panic.

chronic Continuing or recurring over a long period of time.

cortisol A hormone that is created in the adrenal gland and controls the body's stress response.

epidemiology The branch of medicine focused on how diseases and other health factors are spread and potentially controlled.

epigenetic How factors like the environment or behaviors can affect the way genes behave.

equilibrium A state of balance and tranquility.

exacerbate To make worse.

glucocorticoids A class of steroid hormones that helps fight inflammation and supports the immune system. Cortisol is one type of glucocorticoid.

hippocampus Part of the brain that is associated with memory and the regulation of emotions, including anxiety.

mechanism The process by which something takes place.

paternal Something regarding or belonging to a father.

psychoneuroimmunologist A doctor who studies how the nervous system and immune system interact with psychological processes.

resilience The ability to withstand difficulties.

tangible Something that is clearly and definitely real.

FURTHER INFORMATION

"Stress," World Health Organization, February 21, 2023, https://www.who.int/news-room/questions-and-answers/item/stress.

"Stress Symptoms: Effects on Your Body and Behavior," Mayo Clinic, April 10, 2023, https://www.mayoclinic.org/healthy-lifestyle/stress-management/in-depth/stress-symptoms/art-20050987.

Bender, Steven. "Short Naps Can Improve Memory, Increase Productivity, Reduce Stress and Promote a Healthier Heart," The Conversation, August 28, 2023, https://theconversation.com/short-naps-can-improve-memory-increase-productivity-reduce-stress-and-promote-a-healthier-heart-210449.

Fischman, John, and Now Medical Studios. "See How Stress Affects Inflammatory Bowel Disease," *Scientific American*, September 1, 2023, https://www.scientificamerican.com/article/see-how-stress-affects-inflammatory-bowel-disease/.

Jabr, Ferris. "Take That Vacation: Why Time Off Makes You a Better Worker," *Scientific American*, September 1, 2016, https://www.scientificamerican.com/article/take-that-vacation-why-time-off-makes-you-a-better-worker/.

Kahhalé, Isabella, Kelly R. Barry, and Jamie L. Hanson. "Positive Parenting Moderates Associations Between Childhood Stress and Corticolimbic Structure," *PNAS Nexus*, June 6, 2023, https://academic.oup.com/pnasnexus/article/2/6/pgad145/7192947.

Reynolds Lewis, Katherine. "Teen Brains Aged Faster than Normal from Pandemic Stress, Study Says," *Washington Post,* December 1, 2022, https://www.washingtonpost.com/wellness/2022/12/01/pandemic-stress-aging-teen-brains/.

Stix, Gary. "The Neuroscience of True Grit," *Scientific American*, March 1, 2011, https://www.scientificamerican.com/article/the-neuroscience-of-true-grit/.

CITATIONS

1.1 Female Stress—a Faster, Stronger Response by Debra A. Bangasser (November 1, 2016); 1.2 Gray Hair Can Return to Its Original Color—and Stress Is Involved, of Course by Diana Kwon (June 22, 2021); 1.3 The Danger of Stress by Melinda Wenner (August 15, 2008); 1.4 How Short-Term Stress Boosts Immune Systems by Ferris Jabr (August 1, 2013); 1.5 How Stress Messes With Your Gut by Carin Leong, Tanya Lewis, Josh Fischman (July 19, 2023); 1.6 Fact or Fiction?: Stress Causes Gray Hair by Coco Ballantyne (October 24, 2007); 1.7 Does Stress Feed Cancer? by Katie Moisse (April 13, 2010); 1.8 How Dad's Stresses Get Passed Along to Offspring How Dad's Stresses Get Passed Along to Offspring by Esther Landhuis (November 8, 2018); 2.1 Everyday Stress Can Shut Down the Brain's Chief Command Center by Amy Arnsten, Carolyn M. Mazure, Rajita Sinha (April 1, 2012); 2.2 "Stress Hormone" Cortisol Linked to Early Toll on Thinking Ability by Karen Weintraub (October 25, 2018); 2.3 Brain's Dumped DNA May Lead to Stress, Depression by Knvul Sheikh (September 13, 2018); 2.4 Childhood Stress Decreases Size of Brain Regions by Christie Nicholson (August 16, 2014); 2.5 Why Do We Crave Sweets When We're Stressed? by Achim Peters (February 27, 2019); 2.6 Strain on the Brain by Brian Mossop (July 1, 2011); 3.1 How Parents' Trauma Leaves Biological Traces in Children by Rachel Yehuda (July 1, 2022); 3.2 Burned Out by Ulrich Kraft (June 1, 2006); 3.3 The Biggest Psychological Experiment in History Is Running Now by Lydia Denworth (July 1, 2020); 3.4 How Economic Inequality Inflicts Real Biological Harm by Robert M. Sapolsky (November 1, 2018); 3.5 The Stress of Crowds by Alla Katsnelson (September 1, 2011); 3.6 The Myth of Executive Stress by Keith Payne (September 24, 2013); 3.7 How Much Worry about Mass Shootings Is Too Much? by Stephanie Pappas (June 4, 2023); 3.8 News about Racial Violence Harms Black People's Mental Health by Melba Newsome (June 8, 2021); 3.9 Why Do We Panic? by Scott O. Lilienfeld, Hal Arkowitz (October 1, 2008); 4.1 Your Response to Stress Improves as You Grow Older by Daisy Yuhas (December 30, 2022); 4.2 A Single, Quick 'Mindset' Exercise Protects against Adolescent Stress by Lydia Denworth (July 6, 2022); 4.3 The Nature Cure by Jason G. Goldman (September 12, 2019); 4.4 How I Broke the Cycle of Stress by Ashten Duncan (February 28, 2020); 4.5 Enhance Your Resilience by Steven M. Southwick, Dennis S. Charney (July 1, 2013); 4.6 Vision and Breathing May Be the Secrets to Surviving 2020 by Jessica Wapner (November 16, 2020); 4.7 Stress Relief Can Be the Key to Success in School by Clancy Blair (September 1, 2012)

Each author biography was accurate at the time the article was originally published.

INDEX